APPLIED GENETIC ENGINEERING

APPLIED
GENETIC ENGINEERING

Future Trends and Problems

by

Morris A. Levin
Environmental Protection Agency
Washington, DC

George H. Kidd
International Plant Research Institute
San Carlos, CA

Robert H. Zaugg
AgroBiotics Inc.
Baltimore, MD

Jeffrey R. Swarz
AgroBiotics Inc.
Baltimore, MD

np | **NOYES PUBLICATIONS**
Park Ridge, New Jersey, USA

CHEMISTRY

7118 - 6165

Published in the United States of America by
Noyes Publications
Mill Road, Park Ridge, New Jersey 07656

10 9 8 7 6 5 4 3 2 1

Library of Congress Cataloging in Publication Data

Main entry under title:

Applied genetic engineering.

    Bibliography: p.
    Includes index.
    1. Genetic engineering--Industrial applications.
I. Levin, Morris A.
TP248.6.A66 1983   575.1   82-14401
ISBN 0-8155-0925-1

Preface

The book is divided into six parts. Chapters 1 and 2 provide, respectively, historical background for and a description of the techniques utilized by the applied genetics industry. Chapter 3 contains an overview of the biotechnology industry looking back from the year 1990, and analyzes potential hazards summarizing our general understanding of the sources and magnitudes of the potential adverse environmental and health effects of applied genetics. Questions requiring further study are identified, as are possible areas of future concern. As will be apparent, all of the trends or possibilities identified in later chapters will not develop into significant industrial applications. Chapter 3 examines the probabilities of successes based on the interaction of economies, regulating forces and technological input.

The information in Chapter 4 is organized according to industrial sector. The following industries are examined:

- Pharmaceuticals
- Industrial chemicals
- Energy
- Mining
- Pollution and waste management
- Electronics

Chapter 5 contains a description of agricultural applications and trends relevant to biotechnology. Some of the more specialized organisms and techniques are discussed in terms of application, potential and possible hazards. Chapter 6 documents the involvement of and relationship among academic, government, and commercial concerns that have a

v

stake in the applied genetics industry. Both foreign and domestic concerns are included.

The material in this book is based on a series of reports prepared for the Office of Exploratory Research of the United States Environmental Protection Agency. The authors were involved either in preparing or organizing one or more of them. Although the initial reports were reviewed by the Agency, the opinions herein are those of the authors, and do not necessarily reflect the views and policies of the United States Environmental Protection Agency.

Washington, DC   Morris A. Levin
September, 1982

ABOUT THE AUTHORS

Morris A. Levin, Ph.D., is presently a member of the interdisciplinary staff of the Office of Strategic Assessment and Special Studies of the U.S. Environmental Protection Agency, Washington, DC.

George H. Kidd, Ph.D., is the Director of Marketing for the International Plant Research Institute in San Carlos, California. He is an editorial adviser for *Genetic Technology News* and has advised several major corporations on their genetic engineering programs.

Robert H. Zaugg, Ph.D., is presently with Sandoz Incorporated in East Hanover, New Jersey. He was previously Vice-President of AgroBiotics Incorporated, Baltimore, Maryland.

Jeffrey R. Swarz, Ph.D., is presently with Pall Corporation of Glen Cove, New York. He was previously President of AgroBiotics, Baltimore, Maryland.

Contents

1

Introduction

BRIEF HISTORY

The purposeful manipulation of hereditary information in plants and animals by humans, as well as the exploitation of microbial processes, has occurred since the time mankind formed societies. An understanding of the biological nature of these processes has been acquired only recently (i.e., during the past 100 years), and their chemical basis has been unravelled even more recently (in the past 35 years). A variety of terms is now employed to encompass this field of knowledge, including applied genetics, biotechnology, bioengineering, and genetic engineering. While recognizing that these general terms connote subtle differences in scope, we will use them interchangeably in this book. However, certain bioengineering procedures, such as recombinant DNA technology, entail specific activities that require more careful definition. These techniques are described in detail in Chapter 2.

Examples of genetic practices and microbial processes that have ancient origin include the following: alcohol fermentation, cheese production, food crop and domestic animal breeding, crop rotation, and the use of human and animal wastes as fertilizers. The utility of animal and plant breeding and selection was long ago recognized as a controlled method of generating improved strains of vital food crops and hardier domesticated animals. This ancient realization likely arose from the observation that children tended to possess various features characteristic of each of the parents, although the reasons for these similarities were unknown. Alcohol and cheese fermentations were undertaken long before the microbial basis for these processes was recognized. Likewise, occasional planting of fields with leguminous crops, such as soybeans, peas and alfalfa, proved to be a helpful, often crucial, means of replenishing spent soil before it became known that bacteria were responsible for this outcome by virtue of their ability to convert atmospheric nitrogen into usable, chemically reduced forms, such as ammonia. This is the process of nitrogen fixation. And, lastly, ignorance of the role of soil bacteria in recycling human and animal solid wastes did not prevent ancient cultures from employing this rich source of nutrients to improve crop production.

The biological basis of these various processes was recognized beginning in the latter half of the nineteenth century. Two separate findings were essential to the genesis of this understanding. First, during the years 1856 to 1868, an Austrian monk named Gregor Mendel demonstrated in his experiments with peas that numerous observable traits, such as flower and seed colors, are passed from parent to offspring in the form of discrete units of heredity and that each parent supplied independent traits. These revolutionary findings, which were ignored by the scientific community until early in the twentieth century, provide the basis for the gene theory of inheritance, which states that the multitude of traits that constitute an individual organism are expressions of discrete hereditary units, called genes. In higher organisms, these genes are located on chromosomes within the nucleus of each cell. In lower forms of life, such as bacteria, which lack a defined nucleus, the chromosomes nevertheless consist of genes. In all life forms, genes provide the information that determines the make-up of the organism itself, as well as the means whereby traits are extended to the next generation.

The second fundamental discovery that led to an understanding of the biological nature of ancient endeavors in the realm of applied genetics was that of Louis Pasteur. In 1860, he demonstrated that alcohol production from fermentable substrates depended on the presence of viable microorganisms called yeasts. This finding provided the initial example of a living microbe performing a commercially useful process. Today's genetic engineering industry holds the promise that many thousands of commercially useful products and processes will result from applications of recent discoveries in biology that owe their heritage in part to the findings of Mendel and Pasteur.

The chemical basis of genetics was uncovered only recently. Although DNA (deoxyribonucleic acid) was located in cell nuclei in 1869, its role as the bearer of genetic information was not revealed until 1944 by Oswald Avery and co-workers. They demonstrated that pure DNA isolated from virulent pneumococci bacteria was absorbed by a nonvirulent pneumonia strain which was thereupon transfored to the virulent form. Further substantiation of the genetic function of DNA was provided in 1952 by A.D. Hershey and M. Chase, who radioactively labeled both protein and DNA constituents of bacteriophage viruses. (Bacteriophage are simple viruses that infect bacteria; they consist solely of a protein coat surrounding a DNA core.) Infection of susceptible bacteria by these radiolabeled viruses resulted in the finding that viral DNA is necessary and sufficient to mediate the infection. Viral protein is not required.

The above-mentioned studies confirmed the role of DNA as the bearer of genetic information in living systems. It is now well-established that DNA alone serves this purpose in all forms of life, both plants and animals, both primitive and advanced. The information contained within the chemical structure of DNA determines to the full extent the biological nature of the organism (i.e., its appearance and its life functions). (The only exception to the universality of DNA as the genetic material is certain viruses, called retroviruses, that employ ribonucleic acid, or RNA, in this role. Although they constitute an exceedingly small proportion of the total biota on the planet, these viruses are important because they induce malignant tumors in mammals including, probably, humans. Even so, retrovirus RNA is copied into DNA during the process of infection.)

Knowledge of the chemical means whereby DNA maintains and replicates the cell's store of genetic information evolved during the 1950s and 1960s. Many scientific investigators contributed during this time to this advance in understanding, but several steps in particular bear mentioning. In 1953, James D. Watson and Francis Crick proposed a double-helical structure for DNA. This model readily suggested a mechanism whereby DNA could be faithfully reproduced. During the mid-1960s, Arthur Kornberg and associates worked out many of the biochemical details of this replication process. Meanwhile, the genetic code was being broken, most notably by Marshall Nirenberg and colleagues. This code determines how the sequence of chemical constituents in DNA is translated into a specific sequence of amino acids (via a nucleic acid intermediate called messenger RNA or mRNA). Amino acids are the chemical building blocks of proteins which, in turn, provide structural integrity and mediate metabolic activities within every cell of every organism. The steps in the pathway from DNA to protein are diagrammed in Figure 1.1.

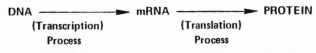

DNA ————————▶ mRNA ————————▶ PROTEIN
         (Transcription)   (Translation)
            Process   Process

DNA IN THE GENES IS TRANSCRIBED INTO MESSENGER
RNA (mRNA) WHICH IS THEN TRANSLATED BY REACTIONS
IN THE CELL INTO PROTEIN.

Figure 1.1: The expression of genetic information in the cell. (Modified from Harsanyi and Karney, Office of Technology Assessment, Report on Biotechnology, OTA HR 132.)

This basic research in molecular biology and genetics paved the way for developments during the 1970s that have given rise to the technology of recombinant DNA. These later achievements and procedures will be detailed in Chapter 2. It must be recognized that the modern field of applied genetics, with all its promise for future benefits to mankind (and its potential dangers), could not exist today but for the numerous accomplishments in basic research in biology and biochemistry over the past several decades, only a few of which are mentioned above.

2

Technology of Applied Genetics

Applied genetics as practiced by ancient societies involved a minimum of human intervention and consisted of little more than allowing nature to take its course. Thus, alcohol and cheese fermentation and the recycling of wastes, processes that we now know to be mediated by microorganisms, were undertaken merely by exposing the appropriate raw materials to the environment, whereupon a transformation of the substrates took place. Controlled animal and plant breeding was implemented by placing prospective parents in proximity to one another. Ancient bioengineering technology, therefore, succeeded by virtue of man's ability to manipulate crudely the biology of his environment.

By contrast, the emergence of modern biotechnology as a scientific discipline that holds enormous potential for benefiting mankind stems from our recently acquired ability to comprehend and manipulate the chemistry of living systems. Thus, the currently popular notion that modern society is embarking on the "Age of Biology" could be slightly rephrased to become the "Age of Biochemistry."

The modern technology of applied genetics encompasses a variety of procedures and processes. Each of these will be dealt with separately in the remainder of this chapter.

RECOMBINANT DNA TECHNIQUES

Recombinant DNA technology refers to the ability to isolate fragments of DNA from separate sources and to splice them together chemically into a functional unit. The DNA fragments can derive from the same organisms or from different organisms in the same species (techniques that have considerable future potential for gene therapy application in humans), but the currently most promising technique involves the joining of DNA segments from disparate species of organisms, such as bacteria and humans. This latter approach has been utilized, for example, in recent efforts to mass-produce human interferon, a drug that may combat viral diseases and cancer.

A review of the recent developments in molecular biology that have led to the emergence of recombinant DNA technology can best be presented by considering those specific laboratory procedures necessary to carry out such experiments. There exist six distinct phases in the process.

(1) Isolation and Purification of DNA–Since DNA exists naturally as a long, fragile, chain-like structure, techniques for gently isolating extended sequences containing intact genes were needed. Such procedures, which include high-speed centrifugation and electrophoresis, were developed during the early 1960s, largely by Julius Marmur and colleagues.

(2) Fragmentation of DNA into Reassociable Segments–This crucial step is mediated by a class of bacterial enzymes, called restriction endonucleases, that introduce widely spaced breaks at specific sites in the DNA chain. The nature of the cuts is such that the separated ends (so-called "sticky ends") can readily reassociate with one another, thereby regenerating the original cleavage site. The rejoining can involve two DNA segments that each derive from different sources, so long as the DNA from each source was clipped into fragments by the same restriction endonuclease. Discovery of these enzymes and elucidation of their physiological role are largely credited to Werner Arber in Switzerland and to Dan Nathans and Hamilton Smith at Johns Hopkins.

(3) Sealing DNA Fragments Together–The rejoining of DNA fragments by way of their sticky ends requires a further step for the full stabilization of the recombined unit. Another enzyme, called polynucleotide ligase or simply ligase, performs this function. The ligase enzymes were discovered independently by a number of investigators, including Malcolm Gefter at MIT and Arthur Kornberg at Stanford.

(4) Replication and Maintenance of Recombinant DNA Molecules– Once DNA fragments have been cut-and-spliced together *in vitro,* a suitable host organism must be found into which the recombinant DNA can be stably incorporated and reproduced. The enteric bacterium, *Escherichia coli,* (or *E. coli*), was the obvious first choice as a host since more is known about the genetics and molecular biology of this microbe than of any other organism. The replication of a DNA segment by *E. coli* requires that the segment contain a specific short sequence of DNA that serves as a signal to the enzymatic machinery inside the cell. This signal, sometimes called the origin of replication, can be found on certain

small, self-replicating loops of DNA, called plasmids, that are commonly found inside bacterial cells. (Plasmids reproduce themselves independently of the major chromosome in bacteria and they are readily transferred between different bacterial strains. In addition to other functions, plasmids are responsible for the resistance to numerous antibiotics that has become a major medical problem in recent years.) Thus, incorporation *in vitro* of the recombinant DNA molecule into a bacterial plasmid, followed by reintroduction of the hybrid plasmid into the bacterial cell, will permit stable replication of the recombinant DNA.

Alternatively, if the recombinant DNA could be incorporated into the major chromosome of the host bacterium, then it would be replicated as part of the chromosome. This is possible through the use of a particular bacteriophage, called lambda, that infects *E. coli.* Upon infection, lambda DNA becomes incorporated into the bacterial chromosome where it replicates along with the host chromosome. Thus, attachment of the recombinant DNA molecule to lambda DNA prior to the infection of *E. coli.* will similarly allow replication of the recombinant DNA.

Both plasmids and lambda bacteriophage are termed vectors owing to this ability to transfer recombinant DNA into suitable hosts for replication. A number of scientists pioneered the effort to demonstrate the usefulness of vectors in gaining expression of exogenous or foreign DNA in *E. coli.,* including Stanley Cohen and Paul Berg at Stanford, and Herb Boyer at the University of California, San Francisco.

There exists a direct method of putting foreign DNA into host bacteria without the need for intact viruses or plasmids. Pure, naked DNA can be absorbed by bacterial cells in a process called transformation. This is the procedure used by Avery and co-workers in 1944 to "transform" nonvirulent pneumococcus strains into virulent bacteria. Some bacterial strains, including *E. coli,* must undergo a simple chemical pretreatment with calcium salts in order to make them amenable to DNA uptake.

(5) Selection of Cells Containing Recombinant DNA–Since only a small percentage of potential host bacteria do in fact acquire recombinant DNA by way of these procedures, it is necessary to perform a selection step. Depending on the type of vector used, it is possible to screen for antibiotic resistance (when the vector is a plasmid containing an antibiotic resistance gene) or to screen for the presence of viable bacteriophage viruses (when lambda is used as the vector). These selection methods give rise to clones of bacterial hosts containing recombinant DNA; that is, each bacterium in the clone is derived from a single progenitor cell that multiplied repeatedly, with exact copies of the cell's DNA having been distributed into each daughter cell. The segment of recombinant DNA contained therein is also replicated; that is, it has been cloned.

(6) Expression of Recombinant DNA into Gene Products (Proteins)–
The recently acquired ability to incorporate exogenous DNA into bacteria, and to have that DNA replicated as though part of the bacterial
genetic complement, is of considerable scientific interest. But commercial applications of this new technology demand that foreign genes im-
planted into bacteria be expressed into the proteins encoded by that
DNA. For example, in order to convert *E. coli* into "factories" capable of
producing human insulin, it is necessary both that the gene for insulin
is stably maintained in the bacteria *and* that the human DNA segment
is transcribed into messenger RNA, then translated into insulin (see
Figures 1.1 and 4.1). As mentioned above, gene replication (maintenance)
is assured by the presence of certain genetic signals. Similarly, the
processes of transcription and translation rely on signals that inform
the cell's enzymatic machinery where to start and where to terminate
each of these processes. All of these various signals must be present at
the appropriate locations in the DNA in order for gene expression by recombinant DNA methodology to be successful.

Once a bacterial cell has been "tricked" into manufacturing a human or other foreign protein, additional problems arise. The bacterium
may recognize insulin as a "foreign" protein and may degrade it before
it can be recovered. If stable, the foreign protein may simply accumulate inside the bacterial cell, necessitating its recovery by breaking
open the cells–a tedious and inefficient process. Ideally, the foreign protein will be excreted out of the host cell into the growth medium from
which it can be readily purified. Clever techniques are now available to
bring this about, and improvements are being made continuously.

One additional roadblock bears mentioning. Many human proteins
possess attachments that consist of sugar molecules. These glycopro-
teins are especially common in blood serum; e.g., interferon as a glyco-
protein, although insulin is not. Bacteria do not possess the machinery
to synthesize or attach sugars to proteins. Although the precise function
of the sugars is unclear, it is probable that they serve a useful, perhaps
crucial, role in maintaining the physiological activity of the protein.
Thus, considerable effort is underway to develop microbial host organisms that *can* attach sugars to proteins. Common brewer's yeast, or *Saccharomyces cerevisiae*, is likely to be the preferred host cell for this purpose. Although it is a single-celled microbe, yeasts belong to the general
class of higher organisms that include humans, namely eukaryotes.
Eukaryotic organisms are classified on the basis of their having a nuclear
membrane surrounding the genetic material within each cell. Bacteria
and certain algae, on the other hand, compose the class of organisms
called prokaryotes (i.e., those lacking a defined nuclear membrane). Although researchers in recombinant DNA have predominately utilized
E. coli as the host organism, there is no doubt that the future com-

mercial success of the technology hinges on the increasing use of eukaryotic hosts such as yeasts and fungi.

A general scheme showing the steps involved in a recombinant DNA
experiment is diagrammed in Figure 2.1.

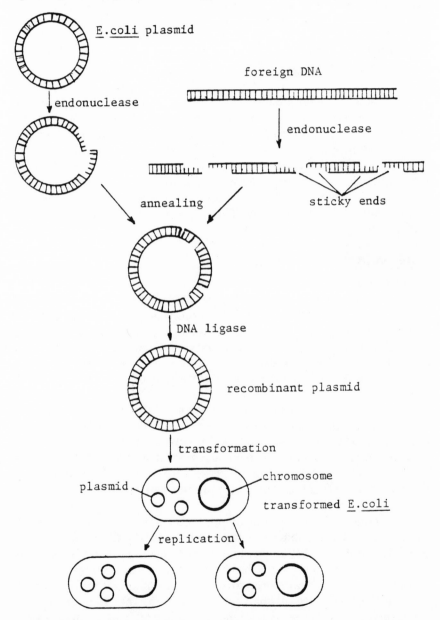

Figure 2.1: Generalized scheme depicting the steps in conducting a
recombinant DNA experiment.

Once engineered, recombinant cells can be, and are saved. Collections of millions of recombinant bacteria, each bacterium containing a different DNA fragment from another organism, have been made in several laboratories in the U.S., as well as in Japan and Europe. These collections are called gene libraries.

GENETIC ALTERATIONS INDUCED BY NONRECOMBINANT DNA PROCEDURES

A number of techniques are currently utilized to induce genetic alterations in cells. The technology of recombinant DNA represents the most recently developed and the most glamorous of these procedures, and it holds the powerful advantage that the outcome of these alterations can be predicted and controlled to a greater extent than with other techniques. Nevertheless, other gene-altering procedures are available, several of which had been in use for many years prior to the advent of recombinant DNA technology. These alternative methods will be described next.

Induced Mutations (Mutagenesis)

The DNA of all living cells is continuously undergoing slight changes in its composition as a consequence of interacting with its external environment. These alterations, called mutations, are thought to be the driving force for the evolution of organisms into new species and, under natural conditions, they occur at a low rate. Under experimental conditions, however, agents that induce genetic mutations, called mutagens, can be administered in order to accelerate greatly the rate of mutation. This is the process of mutagenesis.

A variety of mutagens are used experimentally and commercially to induce mutations. In general, mutagenic agents operate by interfering with the normal cellular processes involved in the repair of DNA. (Healthy cells maintain this enzymatic system for fixing mutations that arise from naturally occurring sources.) Ultraviolet (UV) radiation and chemical agents such as nitrosoguanidine and acridine are the most commonly used mutagens.

The induction of mutations by methods such as these is highly nonspecific; that is, the experimenter cannot control the genetic site at which the mutation will occur. Therefore, following the mutagenic step, it is necessary to conduct a selection for those mutated organisms that possess the desired traits. For this reason, commercial mutagenesis is feasible only with organisms that have a relatively short generation time, such as microorganisms. Since a single bacterial cell will grow to a visible colony within a few days, it is possible to observe the effect of the mutagenic procedure in short order. Moreover, many thousands of such colonies can be screened simultaneously. Nevertheless, a mutagenic procedure was recently described involving plant cells growing in tissue culture. This advance suggests that genetic alterations in plants generally will become feasible by way of induced mutations.

Mutagenesis methods have found widespread utility in the pharmaceutical industry to enhance production of substances from microbial sources. Particular success has been achieved with the antibiotic penicillin, which derives from a strain of mold, and gentamycin, which is produced by a bacterium of the *Streptomyces* species. Unlike recombinant DNA methods, mutagenesis is incapable of endowing the microbe with properties that it does not already possess. That is, no *new* genetic material is introduced; rather, existing capabilities are enhanced.

Cell Fusion Methods

A method whereby one cell type can be endowed with properties of another cell involves fusing those two cells together into a single unit. This procedure is now commonplace in the experimental laboratory and has been applied to a variety of cells from microbes to man and from both plants and animals. The methodology is relatively inexpensive and is essentially the same regardless of the cell type involved. Two technical approaches are in general use.

(1) Monoclonal Antibodies (Hybridomas)–Mammals have evolved a complex internal system of defense against foreign intruders, such as bacteria and viruses. This immune system functions in part by producing proteins called immunoglobulins, or antibodies, that specifically recognize and eliminate these alien invaders. A typical immune response to a bacterium, for example, consists of a variety of different antibody molecules, each capable of recognizing and binding to a specific antigen on the surface of the microbe. Each of these distinct antibody types is manufactured by a clone of antibody-producing cells. These cells are called lymphocytes and since numerous clones of lymphocytes are each reacting to the presence of the bacterium, the response is termed polyclonal.

Antibody preparations (antisera) have long been used to great advantage as diagnostic agents. Immunodiagnostic assays currently comprise approximately one-fourth of all tests performed in the clinical laboratory. Such assays are helpful in rapidly diagnosing bacterial or viral infections and in monitoring drug or hormone levels in blood and urine.

From a physiological standpoint, polyclonal antibody responses to antigens are highly advantageous since they ensure that the individual will effectively repel foreign invaders. But to the clinical chemist, the diversity of antibodies can be bothersome since closely related antigens may not be distinguishable using these conventional antisera. In 1975, a technique was described by Cesar Milstein in Cambridge, England, that permits the generation of monoclonal antibodies; that is, immunoglobulins derived from a single cellular source or a single clone of cells. Such antibody molecules are all chemically equivalent to one another. The technique simply involves mixing antibody-producing cells (lymphocytes) with cells from a type of tumor, called a myeloma, that are themselves derived from lymphocytes. A fusing agent is added that partially dissolves the membrane that surrounds both cell types, thereby permitting contiguous cells to merge together. A common organic polymer, polyethylene glycol, serves as a satisfactory fusing agent. After removal of the fusing substance, the fused cells are grown in tissue culture (see below) and desired clones are identified by a suitable selection procedure. Such a clone combines the qualities of the two contributing cell types: it secretes a specific, monoclonal antibody and grows continuously and rapidly owing to its tumor-like properties. This dual capability is reflected in the term hybridoma, which is applied to a clone of cells secreting monoclonal antibodies. A diagram of the steps involved in generating hybridomas is shown in Figure 2.2.

The initial demonstration of the hybridoma technique and subsequent commercialization of the process have involved cells derived from laboratory mice. The technology was recently extended to the use of human lymphocytes. This advance will soon lead to monoclonal antibodies for *in vivo* diagnostic and therapeutic uses.

(2) Protoplast Fusions and Plant Tissue Culture–The second general class of fusion techniques involves microorganisms or plant cells. These cell types possess a rigid, protective cell wall that surrounds the cell membrane. (Animal cells lack cell walls.) The wall from such cells can be readily removed using enzymes that specifically digest the cellulose-like substance that comprises them. The spherical, membrane-surrounded entity that remains is called a protoplast.

Using methods essentially equivalent to those described for hybridoma production, protoplasts can be fused together generating hybrid cells that exhibit properties in common to both the contributing cell types. Examples of the application of fusion methods to microorganisms include efforts to: (1) improve the antibiotic yield from *Streptomyces* strains: (2) analyze the genetics of brewer's yeasts; and (3) develop hybrid stains of fungi from the *Aspergillus* family to enhance citric acid production.

The application of protoplast technology to plants is a relatively recent development, but one that promises to revolutionize the food, agriculture, and forestry industries, and to have considerable impact on the energy, chemical, and pharmaceutical industries. Scientists are now able to regenerate full grown plants from single cells or proto-

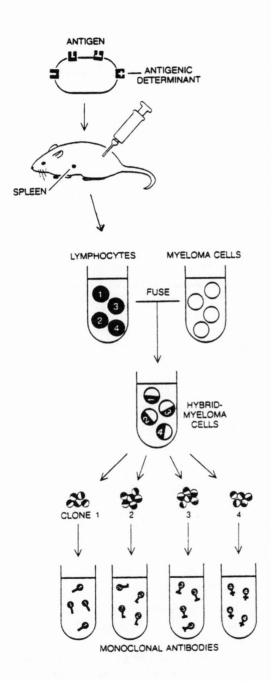

Figure 2.2: Steps in process of generating monoclonal antibodies. (Source: Milstein, C., 1980.)

plasts. So far, only a few species of plant have been successfully cultured in this way, including tobacco, the Douglas fir, and carrots. But rapid advances in this field should soon make available this technology for most plants of commercial interest. This capability will occasion several advantages:

- mass production of clones of identical plants, each having the improved qualities of the original parent;
- rapid growth of the plant in tissue culture to the seedling stage of development, thus shortening generation times;
- a ready-made *in vitro* system for conducting genetic engineering in plants.

A great variety of plants produce chemical compounds that are highly useful to man. These compounds include drugs (such as digitalis, vitamins, steroids, and anticancer agents), rubber and petroleum substitutes. The advent of plant cell and protoplast tissue culture technology makes possible the large-scale fermentation of plant cells in much the same fashion that microorganisms are currently grown in bulk. Useful plant products would be excreted into the growth medium and readily isolated. Future processes of this type will obviate the necessity of devoting large tracts of arable land to cultivation, and production costs should plummet.

The cell fusion procedures described in this section, both for plant and for animal cells, depend greatly for their success on the techniques of *in vitro* cell culture. Scientists have known for some time how to explant cells from particular organisms or tissues and to keep them alive for limited duration under sterile conditions in an incubator. Various nutrient media have been formulated and growth conditions established for a wide variety of plant and animal cells. A serious drawback to the large-scale commercial use of cell culture technology is its high cost, but future widespread industrial application, which seems likely, will introduce economies of scale, and continuing refinements in the techniques should lower costs.

Other Gene-Altering Techniques

Several other methods exist for establishing new genetic material in microbes and in cells of higher plants and animals. Some species of bacteria possess a natural ability to exchange DNA by way of a process called conjugation or plasmids. Extrachromosomal DNA are espcially mobile and are transferred between bacterial strains with considerable ease in some cases. (The ability of bacteria to develop resistance to many types of antibiotics is due to genetic information located on plasmids. Since these plasmids move about so freely, a number of bacterial strains pathogenic to man have become relatively refractory to antibiotic treatment.) Plasmids encoding distinct functions and residing in different bacterial strains can be combined into a single bacterium. Such a "superbug" was created by Chakrabarty, who at that time was working at General Electric. Plasmids from several strains of the species *Pseudomonas,* each capable of degrading a particular constituent of crude petroleum, were combined into a single cell (plasmid associated molecular breeding), enabling the strain that resulted to digest several components of crude oil. This modified bacterial strain became the subject of a controversial patent application, the litigation of which eventually reached the Supreme Court (see Chapter 6).

An alternative to conjugation, which involves bacterium-to-bacterium transfer of DNA, is the process of transduction in which viruses serve as transmitters of genetic material. When a virus infects a cell, normal metabolic activities cease, and processes are undertaken to mass-produce new virus particles. Part of this process involves replicating virus DNA and packaging it into protein shells. Occasionally, small portions of host cell DNA are carried along into the virus shells. After production of sufficient numbers of mature viruses, the host cell bursts, releasing the virus particles to initiate another round of infection. Cells infected in this second round will receive, in addition to viral DNA, the portion of DNA derived from the original host. Scientists have learned how to manipulate these processes so that specific DNA sequences (genes) are transferred, thereby endowing the recipient cells with properties previously inherent only to the initial hosts. The utility of transduction as a means of producing genetic alterations is well established using bacteria and bacterial viruses (bacteriophages). Recently, this general procedure has found application among higher animals. Considerable experimental work is being devoted to performing transduction in primates (e.g., monkeys) using the virus *SV 40* ("SV" strands for "simian virus"). The eventual success of these studies has profound implications for genetic engineering in humans, with the prospect of curing genetic diseases, such as sickle cell anemia.

Genetic engineering in plants promises to be greatly stimulated in the years ahead owing to the existence of a bacterium called *Agrobacterium tumefaciens.* This microbe infects plant cells, giving rise to a plant tumor, called a crown gall. The agrobacterium perpetrates this deed by transmitting to the plant cell a piece of its own genetic material, called T-DNA, which is part of a plasmid, namely the Ti-plasmid (for "tumor-inducing" plasmid). Copies of the T-DNA are incorporated permanently into the plant genes, and the agrobacterium is no longer needed. This instance of naturally occurring recombinant DNA provides a potentially very powerful tool for introducing foreign genes into plants. These, and other specialized techniques and organisms, comprise a large and ever expanding area of research and commercial exploitation. The techniques and organisms are described in detail in Chapter 5. They are highly specialized and are an outgrowth of the agribusiness' general interest in selective breeding for crop improvement.

IMMOBILIZED BIOPROCESSES

Several techniques have evolved in recent years that have managed, to some extent, to exploit cellular biological processes on an industrial scale. These methods generally consist of confining, or immobilizing, intact cells or cellular enzymes within an inert matrix, followed by passage of substrate materials through this bioreactor. Chemical reaction (i.e., bioconversions) then take place that transform the substrate into more useful or less toxic products.

Enzymes are proteins that catalyze the chemical reactions of living cells. Like most proteins, they are relatively unstable and tend to lose their activity when exposed to denaturing conditions, such as heat, extremes of pH and salt concentration, the presence of surface-active agents (detergents) or heavy metals, and so forth. Immobilization procedures generally serve to protect enzymes from denaturation, thereby lengthening their useful lifetimes. A large number of inert support materials have been tested for various applications, including natural and man-made polymers, such as cellulose, starch, polyacrylamide, chitin, polyethylene, glass, and collagen. Enzymes can be either linked securely to the surface of the polymer or entrapped within a porous microcapsule. In order to maximize the reactive surface area, the support matrix can be fashioned into tiny beads or hollow fibers or semi-permeable membranes prior to affixing the enzyme. In all cases, successful operation of the bioreactor depends on maintaining a securely fixed, active preparation that, nevertheless, permits free movement of the substrates and products.

Current and potential applications of this technology are vast and will affect many industrial sectors. A few examples include:

   Chemical industry
   –alkene oxide production from corresponding glycols
   –surfactant production from glycerides
   –hydroxylation of carboxylic acids
   –amino acid synthesis

   Energy industry
   –hydrogen production from water using chloroplast enzymes

–desulfurization of crude oil

–biomass conversions into methanol or ethanol

Medical industry

–production of urocanic acid, a sunscreening agent

–inter-conversion of various penicillin derivatives

–steroid derivitizations

–clinical analysis of blood and urine constituents (e.g., urea and glucose) by electrobiochemical reactions

–synthesis of the antibiotic Gramacidin

Pollution control industry

–conversion of lignocellulosic wastes into useful products, such as glucose

–biodegradation of toxic substances, such as PCB, kepone, dioxin, DDT, phenols

–concentration of toxic heavy metals in waste streams

–air disinfection (e.g., for hospitals) using enzymes that destroy viruses and bacteria

–conversions of whey (waste product from dairy industry) to useful food products

–rotating biological discs for waste water treatment

–fixed-bed bioreactors for on-stream waste management

Food and agriculture industry

–milk coagulation (the first step in cheese production) using the enzyme rennet

–production of high-fructose syrups from starch and cellulose for use as a sugar substitute

–conversion of amino acid isomers to convert the nonnutritious D isomer into the L form

–clarification of fruit juices and wines

Future developments in this area are likely to include improved methods for immobilizing live cells, especially microbes and plant cells. The process of microencapsulation promises to find considerable application here. Each microcapsule can be thought of as a tiny living colony in which cells divide and perform metabolic functions within the confines of the bead. Meanwhile, substrates pass through the beads and are converted into products which flow out of the system uncontaminated by cellular material. Moreover, the biocatalytic system would be self-regenerating since the micro-colonies within each capsule are undergoing continuous turnover; that is, dead cells are always being replaced by live cells. This form of reactivation never occurs with immobilized enzyme systems since isolated enzymes are not capable of self-rejuvenation and, upon inactivation, must be replaced.

Another likely development in the area of immobilized bioprocesses is the increased use of enzymes isolated from thermophilic bacteria. These microbes are remarkably insensitive to high temperatures, even up to 80° or 90° Celsius (water boils at 100°C). Thermophiles can be recovered from hot springs or other similar environments. They owe their heat resistance to having enzymes that are extremely insensitive to heat denaturation. Thus, these enzymes are considerably more stable then comparable enzymes from mesophilic organisms and are ideal for immobilization processes.

FERMENTATION TECHNOLOGY

Commercialization of processes reliant on recombinant DNA or other modern biotechnologies will frequently entail large-scale microbial fermentations. Industrial fermentations have been carried out with great efficiency for many years and have made available at low cost such products as antibiotics, flavoring and coloring agents, amino and organic acids, and vitamins. The expectation that new drugs, such as interferon and human insulin, will soon be mass-produced depends to a large extent on the ability of the fermentation engineers to adapt the appropriate microorganisms for growth in quantities vastly greater than those encountered in the laboratory.

A standard aerobic fermentor consists of a closed, cylindrical vessel equipped with a stirrer and internal baffles to provide agitation, heat exchangers to drain off the considerable heat generated during fermentative growth of the microbial culture, an aerator, and one or more inlets for media sampling and harvesting and exhaust gas removal. A device for rapid steam sterilization of the vessel is essential, as are controls designed to monitor and adjust growth conditions, such as temperature, air flow, pressure, pH, and degree of foaming. Crucial to the overall design is that no foreign microorganisms gain access to the system; all components are sealed to prevent leakage and can be steam-sterilized between batches. The fermentor can be of any convenient volume, ranging up to about 100,000 gallons, in which the media alone would weigh more than 400 tons.

A fermentor designed by Eli Lilly for large-scale growth of recombinant DNA organisms is shown in Figure 2.3. Incorporating design features such as exhaust gas filtration and double agitator seals, this reactor exceeds the safety and containment specifications of typical fermentors and, as such, sets the standard for fermentors designed for use with recombinant DNA organisms.

The fermentor described above carries out a batch fermentation; that is, the media and microorganisms are mixed, the microbes grow for a fixed period (usually one to seven days depending on the organism and the conditions), then the culture is harvested. After cleaning and sterilizing, the fermentor is ready for another batch. Following completion of the batch, the fermentation product must be isolated from the culture system. If the microbe excretes the desired product into the growth medium, as is preferable, then the culture broth must be processed following removal of the microbial population. On the other hand, products that accumulate within the microbe must be recovered by lysing the microorganisms after discarding or recycling the culture liquids.

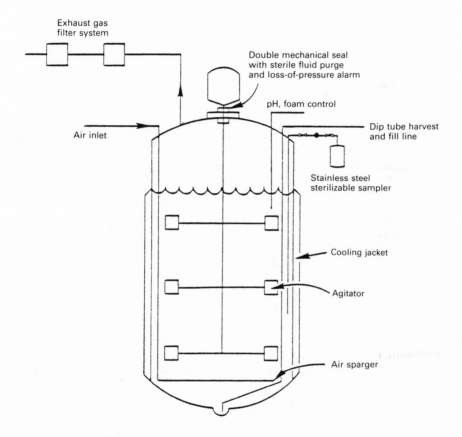

Figure 2.3: Features of a contained fermentor. (Source: Eli Lilly and Co.).

Fermentation technology has advanced in at least two ways in recent years. It is now possible to conduct continuous fermentation in which growth media are added slowly through the growing phase of the microbial culture. At the same time, small portions of the culture are continually removed from the fermentor for processing. The possibility of inadvertent contamination would seem to be greater for this continuous method, but efficiencies are greater since downtimes between batches are eliminated.

A second advance in this technology is called solid phase fermentation. In this process, nutrient media are trickled through a reactor consisting of a solid support matrix to which the microorganisms are steadfastly attached. The microbes continuously excrete the product of interest into the broth which eventually emerges at the bottom of the fermentor. The broth is then processed to isolate the fermentation product. Clearly, solid phase fermentation methods are not applicable to the mass-production of substances that accumulate inside the microorganisms.

GENE THERAPY

Perhaps the most exciting topic in the field of applied genetics and likely the most controversial, is the probability that medical scientists will soon be able to perform genetic engineering in humans. Society in general views this prospect with a mixture of hope and skepticism, and assurances have been given that this capability is many years away. But recent scientific developments suggest that this future may be here quite soon.

Considerable attention was recently directed to the efforts of a team of UCLA scientists, headed by Martin J. Cline, who traveled to Israel and Italy to conduct experiments on human subjects. These experiments were deemed too preliminary to be performed in the United States. Those patients who were treated suffer from a genetic blood disease, called beta-thalassemia, in which production of one of the two protein components of hemoglobin is most negligible. The therapy attempted to insert copies of normal genes for hemoglobin into cells of the bone marrow, where hemoglobin is synthesized. The experiment was given very little chance of success, and, in fact, did not succeed, but the mere fact that it was attempted, added to the fact that a similar experiment succeeded in laboratory animals, hints strongly that some primitive form of gene therapy in humans will shortly be possible (As a result of these experiments, Dr. Cline was reprimanded by the U.S. Department of Health and Human Services and his federal funding was terminated.)

In another recent development, scientists transplanted cell nuclei from early embryos of mice into fertilized eggs isolated from a different mouse strain. After several days in tissue culture, these new embryos were inserted into the uterus of a third mouse. These embryos developed into normal infant mice that were related genetically to the mouse that originally donated the cell nuclei. This outcome has been hailed as the first instance of cloning in mammals. That is, identical offspring were produced by taking cells, or cell nuclei, from a single individual and growing them up into complete, adult organisms. So far, it has not been possible to generate clones from adult donor cells–only cells derived from an early stage of development, such as the embryo, are suitable. But this stumbling block may soon be overcome. If so, and despite the claim by scientists that these experiments are designed only to study gene expression in mammals, then society will be faced with some very sticky ethical issues.

PLANT TECHNIQUES

Plant tissue culture is another form of genetic engineering. Tissue cultures can take several forms including callus, suspension, and single cell cultures; but all cultures are produced by the same basic method. Cells, tissue, or organs are removed from a plant and grown in an artificially controlled environment on a growth medium containing precise quantities of mineral salts, sugar, vitamins, and a combination of plant hormones. This control allows the experimenter to manipulate the plant material to achieve the desired end result.

Perhaps the most remarkable aspect of plant tissue culture is the ability to regenerate whole plants from callus or suspension cells by transfer to a modified culture medium. The modifications primarily responsible for triggering plant regeneration involve the concentrations and kinds of plant hormones used in the medium. Three steps involved in plant tissue culture are as follows:

- Initiate independent single cell growth from plant tissues.

- Grow these undifferentiated cells for several generations in liquid culture medium, and

- Subsequently regenerate whole plants from the cultured cells.

This cycle can be repeated as often as desired.

By changing the content of the growth medium in Step 2 above, the experimenter can select cells with new properties. In this way, tissue culture may provide plants that are disease resistant, stress tolerant, virus free, or nutritionally self-sufficient. Using tissue culture methods, genetic variability in cells can be increased. New breeding systems, that require just a fraction of the developmental time usually needed for conventional breeding, are possible.

3

Tomorrow: Outcome and Problems

A SCENARIO OF THE IMPACT OF GENETIC ENGINEERING IN 1990–A GLOBAL PERSPECTIVE*

Political Situation

The United States (U.S.) and, to a lesser degree, other Organization for Economic Cooperation and Development (OECD) countries have drifted toward policies engendered by a conservative political climate. Confronted with progressively more aggressive global strategies by the U.S.S.R. and other Communist-Bloc countries, and having fully overcome the post-Vietnam psychology, the U.S. has consistently employed a policy of containment and peace-through-strength in its dealings with the Soviet Union and its allies. Backed by conservative factions, the U.S. and its European partners during the 1980s have been successful in increasing military expenditures in their national budgets by 50 percent, including defense-related R&D expenditures. Polarization of the world political situation, leading to the clear identification of hostile as well as friendly nations, has markedly reduced the anxiety and uncertainty that plagued the west in the late 1970s and early 1980s.

Social Economic Situation

Population: The world population has increased at an annual rate just under 2 percent since 1980, and that growth rate is expected to continue through the year 2000. There are now about 5 billion people in the

---

*The reader is reminded that the author of the scenario is in the year 1990 and looking back over the 1980s.

world, and that number is expected to swell to about 6 billion over the next decade. The world's population growth has occurred in the less developed countries (LDCs). This means that of the 5 billion people now in the world, over 3.8 billion live in LDCs. Currently, the average life expectancy is 55 years. Unlike the 1970s, the increase during the 1980s was due more to improved health care than to increased availability of food (see "Food/Agriculture"). However, in certain countries (e.g., India) food production was the overriding factor. Over the past ten years, fertility has dropped by an average of 25 percent, from an average of four children per woman to three.

In addition to rapid population growth, the LDCs have experienced extreme alterations in population distribution during the 1980s. Dramatic movement of people from the countryside to the city continued. Many cities in LDCs became remarkably large and overcrowded. For instance, greater Mexico City, Sao Paulo, and Seoul are much larger than New York City (now about 12 million people). The bulk of the people who have migrated to the cities still live in slums or shantytowns where education, sanitation, and other public services are minimal or nonexistent. The difficulty of maintaining conditions during uncontrolled urban growth occurred alongside worsening conditions in the rural areas of the LDCs. Lack of job opportunities has continued to be severe in these outlying areas, forcing increased migration to the cities.

The large urban populations have resulted in an extreme need for maximum agricultural production in the rural areas. The low income of the average family, however, has created a gap between the felt need for increased agricultural production and the economic demand for this production. Agricultural production even decreased in some countries because governments placed artificial ceilings on food prices in response to urban political pressure. This left farmers unable to recover costs if they were to employ the high energy inputs necessary for maximum production and for expansion to unused land.

Economics: Because population grew rapidly, the worldwide average GNP now stands at about $2400 per capita compared to $2500 in 1975. The rate of growth in worldwide GNP began declining in 1985.

The GNP growth rate in the U.S. and other OECD countries for the past ten years has ranged from 2.5 to 3.0 percent annually, so that the per capita GNP now averages more than $9500. The inflation-recession cycles are still evident, but through the 1980s the OECD countries maintained better control than they did during the 1970s. The U.S. inflation rate was slightly over 8 percent last year (i.e., 1989), and its GNP grew 2.7 percent.

By contrast, the LDCs have witnessed economies expanding at near 6 percent annually over the past decade. However, per capita GNP in the LDCs still averages less than $450, although indicators do point to some progress in economic growth. Nevertheless, economic disparities between wealthy and poor nations continue.

Technological Situation

Global R&D Expenditures: The total R&D expenditure is still greater in the U.S. than in any OECD country. However, the ratio of R&D expenditures to GNP in the U.S. has declined consistently since the 1960s and stood at 2.0 percent by the middle of the 1980s. At the same time, the R&D to GNP ratio for West Germany and Japan has surpassed that of the U.S. Throughout the 1970s and early 1980s the general pattern within the OECD countries, except for the U.S., was to shift away from military to civilian R&D. Recently, there has been an observable and uniform reversal of this trend with almost all OECD governments now expending larger portions of their R&D funds for military related applications. The only nonmilitary research areas that have received special attention in recent years in the U.S. are energy development and conversion, human health, and the environment.

The level of R&D expenditure by industry in the individual OECD countries varied with the performance of the respective economies. Japan and West Germany have enjoyed healthy and growing economies and consequent steady growth in basic, applied and development R&D expenditures by their industries. Even though the economic growth has been only moderate and the inflationary-recessionary cycles have abated in the U.S., industry has become more selective and cautious in its R&D allocations. Expenditure for basic and applied research by U.S. industry still lags behind the U.S. Government's. However, several technologies in the OECD countries have benefited from exceptionally high industrial investment in both basic and applied research activities. Energy-related technologies, chemical and allied technologies, drugs and medicines, food technologies, and electrical and communication technologies are among the prime examples.

Genetic Engineering in General: Against the foregoing political and economic background, genetic engineering had some success in the 1980s, but fell short of the early promises that attributed broad applications of panacean proportions to the technology.

The expectations in the early 1980s about the potential of genetic engineering propelled a large number of high technology industries, such as the chemical industry, into investing in basic and applied research in genetics. Genetic engineering research programs were also begun by several academic and nonprofit research institutions in the U.S. and several other OECD countries, such as Germany and France. Some countries, such as Japan, which by the early 1980s already had in place an infrastructure for developmental research in fermentation aimed at specific medicinal and fuel application, broadened their efforts toward many end-uses. This effort was reinforced by the increased exchange of scientific and technological information and products among the OECD countries. OECD countries have slowly moved toward a course of "less protectionism" in trade, science, and technology among themselves, in part as a countervailing policy against actions of the U.S.S.R.

This increasing international cooperation among the OECD countries created employment and economic opportunities for many biological scientists. Up through the first quarter of the decade, the opportunities were mostly reserved for physical scientists, and more specifically for those who were active in research on energy engineering, medicine, and national defense. Industrial organizations with in-house R&D activities in genetic engineering, profit-making and nonprofit R&D institutions, and organizations specializing in genetic engineering created a highly competitive employment market for scientists with advanced degrees in the biological sciences. The demand for specialized scientists and engineers in genetic engineering was further heightened by the short supply which resulted from the overcommitment of resources in the late 1970s and early 1980s to postgraduate programs other than life sciences. For example, the concern over the inadequate supply of scientists and engineers in such energy-related fields as mining, petroleum engineering and chemical engineering attracted considerable resources in the U.S. Although most institutions of higher learning in the OECD countries had made the adjustment by the mid-1980s and more doctoral-level biological scientists were entering the market, most of the pioneering research was still being carried out in a few organizations and by experienced scientists whose number was less than the growing demand for their know-how.

In the U.S., federal and industrial R&D funding for genetic engineering has grown to $200 and $500 million annually, respectively. Over 180 industrial organizations are now performing genetic engineering R&D. The recombinant DNA products market alone has grown to $3 billion.

Because of the inadequate supply of highly trained scientists needed for the pioneering work in genetic engineering and because of unforeseen experimental problems, obtaining commercial results from the work carried out in the development and application phases has become inhibitively prolonged and costly. For instance, interferon production via recombinant bacteria has greatly slowed in the mid-1980s because of successful production from animal cells.

During the 1980s, applications of genetic engineering technology encountered stiff competition from newly developed organic chemical technology, new drug formulation and delivery technologies, and a resurgent interest in conventional plant and animal breeding. Even though genetic engineering has made moderate progress overall, replacement of traditional industrial processes (e.g., the Haber-Bosch process for producing ammonium fertilizers) are not being considered.

Agriculture and Genetic Engineering: In addition, crop improvement programs centered on overcoming problems arising from the increased pressures put on soils throughout the world. In both industrialized countries and LDCs there has been a spread of desert-like conditions in the drier regions and heavy erosion in the more humid regions. The increased destruction of vegetation cover, the loss in yield, and the decrease in the availability and fertility of soil have mainly been caused by overgrazing, destructive cropping practices, urban encroachment, and use of plants for fuel.

The impetus for the development of genetic engineering in agriculture over the 1980s rose primarily from destructive cropping practices such as irrigation and pesticide application. The loss of irrigated lands over the decades has been significant. Today (1990) about 18 percent of the world's arable 125 million acres is being irrigated. Productivity on about 60 percent of the world's irrigated land has decreased because of increased salinity, alkalinity, and waterlogging. Irrigation's negative impacts on productivity exist in both industrialized countries and LDCs. The extensive use of pesticides into the late 1980s, which sustained the rise in crop yields, resulted in increased resistance of most pests to multiple pesticides and in the destruction of pest-predator populations.

Even with the initial impact of the major seed companies and oil companies entering plant genetic engineering, the application of genetic engineering to agriculture has fallen short of expectations. Early proponents of short-term benefits to agriculture minimized the problem of techniques still to be developed. Even now, no commercially available seed stock contains a new crop gene introduced by genetic engineering. Corn with the capacity to fix nitrogen is closest to commercialization, another six years being required to test the stability of the trait and to produce hybrid seed in quantity. Delays in producing nitrogen-fixing corn have arisen because most strains that fix nitrogen did not retain other attributes judged essential by farmers (e.g., high yield, pest resistance). Other plants, mainly vegetable crops, have been engineered to be salt-tolerant, herbicide-resistant, etc., but commercial seeds with those traits are not yet available.

Early in the 1980s, the possibility of growing isolated plant cells in culture and the subsequent regeneration of whole, seed-bearing, competent plants offered promise in breeding new crop varieties. For instance, cell lines of corn and wheat resistant to specific toxins from microbial pathogens were developed, but the whole plants derived from these cell lines possessed poor grain quality. In any event, conventional wheat and corn breeding have already provided the world with many strains resistant to toxin-producing pathogens. A number of plants have been examined for use as sources of chemical industry products.

Several other factors limited the application of genetic engineering to agriculture. First, funding for basic research in the plant sciences decreased despite world food pressures, because of emphasis on military research. Our data base for scientific experimentation with photosynthetic plants is still more meager than that for health and human research. Second, concern over the impact of using recombinant organisms (including plants) in the environment slowed research in the early 1980s. Third, genetic engineering was aimed at increasing the uniformity of crops. Unfortunately, large monocultures of genetically identical crops increase the risk of catastrophic loss from insect attacks or crop epidemics. The corn blight that struck the U.S. corn belt in 1970 was a clear illustration of the vulnerability of genetically identical monocultures. The possibility of other such epidemics was constantly being raised during the 1980s, thereby impeding research on genetic engineering of plants.

The 1980s further confirmed the idea that genetic engineering must function only as an adjunct to, rather than as a replacement for, conventional plant breeding. Through the 1980s conventional plant breeding has been successful and safe to the environment. Even today the limits to improvement of the major grain crops through conventional plant breeding have not yet been attained. For example, in 1930 average U.S. corn yield was 20 bushels/acre (bu/a), in 1975 86 bu/a, and in 1990 105 bu/a. In another instance, wheat productivity was 14 bu/a in 1930, 30 bu/a in 1975, and 38 bu/a in 1990. The productivity of corn and wheat are still increasing, although at an ever slower rate. The increased yields were due to conventional breeding of superior stock as well as to increased use of energy-intensive factors (fertilizers, pesticides, harvesting equipment).

In addition to the research thrusts in plant improvement that have been on-going for several years, new problems are beginning to spawn new areas of investigation. The versatility, proven record of success, comparatively low technology needs, and consequently moderate cost of conventional plant breeding have resulted in the lead role for both the old and the new research efforts being in the hands of plant breeders. Genetic engineering technology has become, and will continue to be, a tool of last resort for increasing the genetic diversity of a species. It is used by the breeders to provide the occasional important breakthrough they could not otherwise expect.

The steadily worsening impact of chemicals on the environment has resulted in several major research thrusts in plant improvement. Two approaches are being taken to deal with soil contamination with heavy metals, such as sewage sludge, with certain landfill areas, and with reclaimed areas formed from dredgings from rivers and lakes. A variety of plants that take up heavy metals from the soil are being developed for growing on contaminated soil to decrease its metal content. Some are grain-producing crops which do not translocate metals to the grain. The grain will then be fed to animals, and the remaining biomass combusted for energy with the ashes going to hazardous waste areas. Other plants that do not take up specific metals are being developed so that metal-free crops can be grown on contaminated land but used freely in normal commerce. The need for plants resistant to air pollutants is increasing, both because larger geographical areas are affected by pollutants and because higher maximum concentrations are now encountered in certain localities. Certain especially sensitive crops are being developed for increased resistance to ultraviolet radiation in anticipation of continued depletion of the ozone barrier in the atmosphere.

The energy crisis has crippled the economies of many LDCs so that the use of fertilizers and disease-control chemicals has considerably diminished. The resulting impact is heightened because in most tropical settings there are great pressures from pests and rapid leaching of nutrients. One new thrust that is arising in response to this situation is designed to make increased use of mycorrhizae-plant associations. Although such associations are common to most plants, not all mycorrhizae are equally efficient in increasing nutrient availability in poor soils. Genetic engineering techniques are being used to attempt the transfer of appropriate genes from species hosting the most beneficial mycorrhizae to other plants to enable them to form similar symbiotic associations. Attempts are being made to transfer genes coding for rotenone, pyrethrum, and other natural insecticides to a variety of plants to give them a natural resistance to insects. Genes responsible for the ability of certain plants to secrete phytotoxic substances from the roots and for resistance of the host plant to those substances may soon be transferred to create herbicide-producing crops. The potential of these developments for reducing energy requirements is beginning to attract considerable attention.

The depletion of forests in nations where 90 percent of the families rely on charcoal as fuel for cooking has resulted in an extreme crisis, with charcoal as well as food now being sent by relief agencies. Research is now on a crash basis to develop varieties of local staples that require less cooking. The pivotal role of pulses as the major protein source in many countries has resulted in special attention being given to them. Efforts to reduce their exceptionally long cooking time are still at the basic research level. Equally important are efforts to develop varieties that lack the antinutritional and toxic substances that require long cooking times to inactivate. Countless edible plants that have scarcely been utilized until now are being screened as potential new food sources that can be consumed raw or that can be prepared with minimal cooking.

Food/Agriculture: Over the decade of the 1980s, fuel and food needs around the world became the most pressing human problems.

Through the 1980s, world food production increased at an average rate of slightly over 2 percent per year, a rate of increase which is expected to continue through the year 2000. So, estimates indicate that by 2000 food production will be 80 percent greater than it was in the base year 1970.

However, maintenance of these high yields depends on the continued use of energy-intensive inputs like nitrogen fertilizer, pesticides, herbicides, and irrigation. Energy-intensive production methods have continued and will continue to aggravate further fuel availability and oil prices. Intensive agricultural practices and the use of more marginal land have greatly enhanced desertification. Thirty million acres of arable lands were lost during the 1980s. Most good land is already under cultivation, and from 1990 to 2000 land under productive cultivation will increase only about 4 percent (mostly due to the annexation of marginal, sloping lands and to the clearing of forested land). So, by 2000, one acre of land should supply food for 10 people, whereas in 1970 one acre fed about 6 people.

During the 1980s world food production barely kept ahead of the world population; the absolute price of food increased 55 percent. This increase was a direct result of increased petroleum dependence, and the shunting of food/feed reserves (corn grain) to the manufacture of liquid fuels, although by the late 1980s grain-based alcohol production was beginning to be supplanted by coal-to-methanol processes. In the U.S., the cost of producing crops increased about 12 percent per year through the 1980s. Production-cost increases in the LDCs were somewhat lower but still fiscally burdensome to those countries. Adding to the rapidly increasing food production costs was the doubling of food distribution costs (also petroleum dependent) during the decade.

Per capita food consumption in industrialized nations has increased on the average of 15 percent since 1980 and is projected to increase another 15 percent by 2000. Over the past 10 years, per capita food consumption increased about 20 percent in Japan, Eastern Europe, and the U.S.S.R., and about 10 percent in the U.S. and the LDCs. This relatively small overall increase in consumption in the LDCs is misleading because there were enormous variations in consumption among nations and regions. Similar consumption increases in the U.S. and the LDCs have occurred for different reasons. For example, the U.S. experienced an average population growth (about 1 to 2 percent of the world growth) while the LDCs saw rapidly increasing populations but relatively low agricultural productivity. The number of malnourished people in the LDCs now stands at almost 2 billion.

Energy: The energy picture has become increasingly gloomy. Over the last decade there were large increases in demand for all commercial energy resources. The world energy demand is currently 390 quads. Oil is still the world's leading energy source, providing 45 percent of the world's energy supply. The average price of a barrel of crude oil has reached $65.00 (1980 dollars). In view of the increases in demand, less than adequate success in conservation, and development of alternative energy sources by the western countries, the $65 per barrel seems considerably low. Two reasons can be cited for this relatively moderate price. First during early part of the 1980s, the moderate elements within OPEC were successful in consistently dissuading the more radical OPEC countries from raising oil prices in the manner of the 1970s. The winning argument has been the fear of economic collapse in the western countries and its consequences for the oil-producing nations. Second, a succession of inter and intra political upheavals within and between oil-producing countries in the middle-East contributed to disunity within OPEC ranks, thus preventing OPEC from acting in unison on the oil price issue.

Although the price of oil is considered manageable by the western economies the oil-producing countries in the middle-East have proved to be unreliable sources of supply. Constant interruption of supply due to internal political turmoils and between countries squabbles have at times thrown the western world into shocks reminiscent of the oil shortages in the 1970s.

The per capita consumption of energy has increased about 70 percent in industrialized countries other than the U.S. The U.S. and LDCs both witnessed a 30 percent increase in energy consumption over the 1980s. In actual values annual per capita consumption of energy is about 400 million Btu in the U.S. compared to about 15 million Btu in developing nations.

The short supply of wood for fuel has created a serious problem for the LDCs. Over the 1980s the demand for wood increased 25 percent, but the availability of wood decreased about 75 percent. There are virtually no forests left in sub-Saharan and central Africa, in southern Asia, and in central Caribbean and Central American countries. The poor burn only crop residues and dung. Extensive cutting of forests in Southeast Asia and other Asian and African countries has also destroyed water supplies by altering forested mountain watersheds.

Energy/Food: This food-energy relationship has resulted in a broadened range of applied genetics to the energy industry far wider than anticipated. Development of systems for ethanol production from biomass for use in gasohol proceeded apace, especially in petroleum-poor areas like Brazil, but the economics of this process and the energy savings incurred remained unfavorable until the end of the decade. In addition, several long-range projects have been instituted to provide sources of energy.

Production of hydrocarbon substances from higher plants has become economically feasible using plant cells manipulated to grow in massive cultivators, akin to microbial fermentors, in which excreted hydrocarbons are continuously collected.

Biological solar batteries are replacing panels of silicon solar cells. The biological battery operates via a direct conversion of sunlight into electricity (i.e., a current of electrons) that is generated during photosynthesis. *Rhodospirillum rubrum* is being used as the living solar cell. Photosynthetic blue-green algae, which utilize carbon dioxide and nitrogen directly from air are being considered as alternates.

Ethanol production has become more efficient through use of microorganisms other than common yeasts (e.g., *Saccharomyces cerevisiae,* or brewer's yeast). *Zymomonas mobilis* carries out alcoholic fermentation two to three times faster. This bacterium originally employed to make tequila is now widely used.

Acidophilic, iron-oxidizing *Thiobacilli* (commonly used in mineral leaching operations) have proven useful in oil shale and coal conversion processes. The bacteria will mobilize the inorganic mineral content of the shale or coal without affecting the hydrocarbon content of the material. The porous zones that this process generates *in situ* assists in subsequent retorting or gasification schemes.

Minerals Industry: Spurred on by a desire for "Mineral Independence" the researchers in the U.S. investigated the biochemistry and genetics of microorganisms known to dissolve minerals and either concentrate or excrete the mobilized ions ("Leaching bacteria"; e.g., *Thiobacilli).* Initially all known strains of leaching bacteria were aerobic; that is, they required oxygen. However, the essentially oxygen-free conditions existing in the center of huge slag heaps of low-grade ore led to the engineering of anaerobic strains. Development of improved thermophilic leaching bacteria is being attempted to counter the heat generated within ore dumps. Water problems, both quantity available and contamination (acidification) have been encountered. Nevertheless, steadily increasing percentages of the U.S. demand for copper and iron are being produced by these techniques.

In addition, the use of bacteria to concentrate metals from waste streams (as much as 15% by weight) has decreased reliance on imports via increased output from recycling efforts.

Pollution Control Industry: Emphasis on research into the biochemistry of obligate anaerobic and facultative organisms has paid off in a myriad of new products and techniques resulting in less noxious waste emission and degradation of existing hazardous toxic waste products. Bioengineers have concentrated on developing organisms which can treat specific compounds and have installed bioreactors in waste lines where they can be most effective by treating waste from a single source. Complex mixtures of waste material are being treated as in the past by adding genetically engineered mixtures of microorganisms to sewage plants or lagoons. The organisms used have, in general, been selected from isolates derived from the soil/water samples from the site involved.

Biomedical: It is in this field that applied genetics has made the most dramatic, and most controversial impacts. The prospect of genetic engineering in humans raised deeply personal ethical questions that are not of concern to applications of biotechnology to other commercial sectors. Even now, however, specific developments are difficult to predict; often the most significant applications are not even conceived of even a few years in advance.

However, recombinant DNA methods do include the ability to transfer human genetic material into bacteria. The capability depends on certain bacterial vectors, usually plasmids or viruses, that carry foreign DNA into the host. Similarly, transfer of human DNA into other human cells or tissues is feasible through use of appropriate vectors that mediate the exchange. Such vectors, namely mammalian viruses have been developed. Their availability will facilitate genetic engineering in humans. More serious technical questions still stand in the way of medical application, however. For example, which human genes should be transferred in order to treat which disease, and how are those genes isolated? What steps are required to establish those new genes in the recipient individual? Apart from technical obstacles, unresolved political and ethical issues pertaining to experimentation in humans still stalls widespread application of this technology.

More direct applications, such as production of antibiotics from plant tissue cultures, development of new antibiotics and anticancer drugs and diagnostic procedures have become routine.

Monoclonal antibodies ("hybridomas") have been produced using human as well as mouse tissue. Such antibodies have multiple drug uses: as antidotes for acute bacterial or viral infections; as agents for localizing and treating inaccessible tumors; as an adjunct to tissue transplantation to prolong graft survival; as safe contraceptive agents.

There has been a resurgence in the search for natural drug-like substances produced by plants and sea creatures. A variety of powerful drugs (e.g., digitalis, morphine, vincristine/vinblastine cancer drugs, and many steroids) were first isolated from plants. Modern drugs based on these compounds are now chemically synthesized. There exist numerous natural products that may serve as useful drugs but which occur in such limited quantities or which are so difficult to synthesize that commercial development is unlikely. Applied genetics will soon permit mass production of these substances by genetic manipulation of the organisms that produce them (Table 3.1).

Table 3.1: Examples of Pharmacologically Active Products Isolated from Microorganisms

Activity	Product	Producing Strain
Anticoagulant	Phialocin	*Phialocephala repens*
Antidepressant	1,3-Diphenethylurea	*Streptomyces sp.*
Anthelmintic	Avermectin	*Streptomyces avermitilus*
Antilipidemic	Ascofuranone	*Ascochyta viciae*
Antipernicious anemia	Vitamin B_{12}	*Streptomyces griseus*
Coronary vasodilator	Naematolin	*Naematoloma fasciculare*
Detoxicant	Detoxin	*Streptomyces caespitosus*
DNA transformation inhibitor	Antraformin	*Streptomyces sp.*
Esterogenic	Zearalenone	*Gibberella zeae*
Food pigment	Monascin	*Monascus sp.*
Herbicide	Herbicidin	*Streptomyces saganonensis*
Hypotensive	Fusaric acid	*Fusarium sp.*
Immune enhancer	N-acetylmuramyl tripeptide	*Bacillus Cereus*
Insecticide	Piericidin	*Streptomyces mobaraensis*
Miticide	Tetranactin	*Streptomyces aureus*
Plant hormone	Gibberellic acid	*Gibberella fujikuroi*
Salivation inducer	Slaframine	*Rhizoctonia leguminicola*
Serotonin antagonist	HO_{2135}	*Streptomyces griseus*

Source: Woodruff, H.B. (1980), *Science,* 208:1228

Interferon is not a single substance, but exists in multiple forms (numbering at least eight so far). The physiological role of each of these interferons has yet to be unraveled, but a better understanding of this biological system will lead to a wide variety of new drugs for treating specific viral diseases and cancer. Initial methods of production relied on tissue culture techniques, but microorganisms which excrete the molecule are now the production method of choice.

Chemical Industry: The ease with which applied genetics has been integrated into the pharmaceutical sector was a result of that industry's predisposition towards the biological sciences. On the other hand, the chemical industry, which depended largely on the technical disciplines of physical and organic chemistry and chemical engineering for its commercial foundation, lagged in applying the newly available procedures. Although recent decades have seen remarkable advances in the mass production of industrial chemicals that have benefited society in numerous ways, agricultural chemicals that have improved food production, synthetic fibers that have revolutionized the clothing industry, and plastics that influence our lives in countless ways, traditionally the chemical industry had not involved itself with biological processes. Only within the past few years have chemical firms, such as Allied, Dow, DuPont and Monsanto, undertaken major programs to examine biotechnology as a way of doing business in the future.

Despite increased shortages of crude petroleum, the fact that the industry still uses only 7 percent of the total oil consumption (unchanged from 1979) has permitted continued production of bulk chemicals (products whose production capacity is measured in millions of tons) to continue to be produced via traditional synthetic routes. This state of affairs is expected to continue for at least another decade.

However, a significant role for applied genetics in the chemical industry has been in the manufacture of high-priced specialty chemical and in synthesizing new chemicals that have no practical alternative route. Immobilized enzymes have been employed as highly specific catalysts for performing discrete chemical steps in a synthetic route. Microorganisms that express the desired enzyme activity are being used directly. Microbes are now being sought to carry out chemical transformations otherwise requiring large inputs of energy, such as hydrogenation, amidation, etc.

The production of proteins and hormones (using RDNA technology) has doubled in the past 10 years and a 50 percent increase is anticipated by 2000. An even more dramatic shift has occurred in peptide production (up 200X) with a further doubling expected by 2000.

Electronic Industry: The most recent industry to be affected by the biotechnological revolution is electronics. A combination of imagination and mechanical engineering offers the possibility of reduction in computer size (factors of 100-1000) with concomitant reduced power utilization and heat output. The possibilities resulted from utilization of the ability to place a monomolecular strip of protein containing a single amino acid (lysine) on a silicon chip. Coupled with standard techniques for depositing silver on a protein, this procedure provided a means of conducting current which offered little resistance and negligible heat production. Iron atoms with varying values function as "molecular switches." These concepts, developed in 1980-1983 are now being designed into prototype ultrasmall computers.

Technological Growth: The regulations concerning genetic engineering applications during the 1980s fell into five classes. These were:

Research practices and procedures

Types and species of plants

Types of microorganisms

Industrial scale-up and field testing

Proprietary issues, e.g., Plant Protection Act

The areas that were impacted due to the enforcement of regulations are:

Investment

Industrial R&D expenditure

Basic and applied research and technical developments

Two sectors have been concerned with compliance with these regulations.

Academic sector

Industrial sector

The investment inertia that was created in the early 1980s due to expectations for high long-term returns, has especially been sustained throughout the decade. However, there have been short periods when investors reacted cautiously to industrial organizations whose programs of R&D in genetic engineering were highly susceptible to regulatory pressures. When an industrial organization's R&D programs required reconciliation with existing regulations, generation of capital became a major problem. The investors simply would not commit their money until the regulatory issues were resolved.

Among the five areas of regulation, none has had more impact upon investment than the regulations related to the industrial scale-up and field testing. As the last stage in events prior to full commercialization, scale-up and field testing represent the point at which most of the developmental expenditures are considered "sunk" cost. Unless products or processes are forthcoming, the incurred costs cannot be recovered, and profit remains unrealized. From the investors' point of view, delays due to regulatory considerations mean increased risk of return on capital. For the organization, it means a tie-up of expenditure and the slowdown of cash flow and profit. Consequently, when compliance with regulations aimed at scale-up and field testing was involved, investors and companies were discouraged from committing themselves, or they acted cautiously in commitment of capital and organization effort.

Compliance with regulations in the area of research practices and procedures has had different degrees of impact upon investment in R&D. In industry, the enforcement of research procedures and related regulations has progressively resulted in greater R&D costs. The increase in cost has been most noticeable for the experiments requiring P3 and P4 containment laboratories. These regulations have meant more elaborate and sophisticated laboratory construction. Capital-rich industrial organizations have been able to proceed with the R&D. However, smaller organizations engaged in recombinant DNA research have been forced to limit their investigations as dictated by their containment capabilities. Universities, which are traditionally bounded by the availability of funds, have been impacted more severely. A number of universities have been forced to abandon some of their genetic engineering research activities since 1985. On the other hand, academic institutions with historically strong financial ties have successfully complied with the containment requirements. As a consequence, a handful of universities have attracted most of the scientists working in agricultural genetic engineering.

Another consequence of the strict enforcement of research practice regulations has been increased direct labor cost both in the industrial and academic environments. For example, in most of the universities clerical staffs are employed to handle the paperwork and communications related to the regulatory functions. Compliance requirements have also forced the bench researchers at the universities to spend more time not directly related to experimentation. Although this factor has not actually prolonged the research time, it has become a source of complaint and frustration for academic scientists. In industry, increase in the scale of organizational structure has been partly due to regulations in research practice. Industry, unlike universities, has successfully shielded its scientific staff from devoting time to compliance matters. This has been done by increasing the number of organizational levels and specialized personnel.

Regulations on species of plants and microorganisms have dramatically impacted basic and applied research. The effects of these regulations have produced two major consequences. The first, and most obvious, has been limitations upon the scope of research, thus limiting applications of new plants and processes. This effect has been most severe on species of crop plants with large commercial potential. Industry has reacted to these regulations by simply shifting its R&D emphasis to the unregulated areas, without any noticeable loss of enthusiasm for potential commercial payoffs or without reduction of expenditure. Investment has been directed towards commercial ventures whose underlying R&D programs and production processes utilize unregulated plant species and microorganisms.

A significant event in the 1980s has been the development of relatively undisputable and clear rules for identifying the characteristics of genetically engineered plants to be patented. This has given added impetus and incentive to industrial concerns. The perceived commercial potentials have frequently accelerated investment in companies whose proprietary potential for new plants seemed attractive. Although the impact of genetic engineering in development of new plants has been less than anticipated and only a limited number of patents issued, the financial community has acted more on expectations than on actual results.

An argument frequently heard in the early 1980s was that patents may restrain the free exchange of scientific information. This contention has neither been proved or refuted by the experiences of the last few years. However, as a result of patent issues, the scientists are now more inclined to discuss their techniques because the threat of duplication of processes is prohibited by law.

In summary, the regulatory impacts in the 1980s have ranged from negligible to moderate. The regulations have not dampened investment, generally. They have increased the cost of R&D but not to a prohibitive degree (at least for industry). The academic sector, however, has not been able to adapt as easily as industry. Industry has pushed ahead within the regulatory framework. On the other hand, the university scientists, who historically carry out the basic research, have produced less results both in terms of scale and scope.

HAZARDS: REAL/IMAGINED

The range of potential health and environmental hazards posed by applied genetics will vary depending on the industrial setting. In the pharmaceutical and chemical industries, for example, processes involving genetically engineered microorganisms are likely to be contained within closed reactors or fermentors. Many applications of biotechnology in the mining and pollution control industries, on the other hand, foresee deliberate release of microbes into specific open environments. These two general modes of operation clearly impose different risks on (1) the health of the workers involved and of the surrounding community and on (2) the local ecology. Some of these risks are simply those associated with industrial scale efforts in general; others (more problematical and less easy to define) may be associated with recombinant technology per se.

Pharmaceutical Industry

The application of biotechnology in the pharmaceutical industry gives rise to potential hazards at several levels of activity not all peculiar to recombinant technology:

> The research laboratory, where scientist and technical personnel engage in the initial stages of development of new drugs or therapeutic regimens. A potential risk arising from the creation of new microbial strains via recombinant DNA techniques, for example, will be first experienced by laboratory personnel. The specific hazards involved are mitigated, however, by the high level of personnel training in general laboratory safety and by the relatively small quantities of material encountered in the laboratory setting.

> The production facility, where a large-scale manufacturing, product isolation, and packaging processes are undertaken. The drug industry has amassed considerable experience in the safe operation of huge fermentation facilities. There remains the potential risk, however, of exposing the work place (and to a lesser extent, the surrounding community) to aerosols containing viable microorganisms. Although their health is monitored quite closely, production workers are less able than are highly trained lab personnel to recognize the symptoms of microbial infection. Certainly, individuals residing in the surrounding community are generally unqualified to appreciate the risks posed by these activities.

> The end users, including medical personnel and patients, of drugs manufactured through applied genetics. Risks here are minimized by (1) enforcement of existing government regulations pertaining to the introduction of new drugs and biologicals and by (2) strict control of product quality by the manufacturer.

Microbiology laboratories have been examined since the turn of the century as sources of bacterial and viral infections. A recent survey (see R.M. Pike, 1976) summarizes nearly 4000 lab-associated infections dating back to the early 1950s. The most common bacterial and viral diseases reported among lab workers were brucellosis, typhoid, tularemia, and hepatitis. However, fewer than 20 percent of these infections could be associated with a known laboratory accident of any kind. (The lab practices most frequently giving rise to infections are mouth pipetting and the use of needles and syringes.) Although the incidence of such infections among lab workers is 5- to 10-fold higher than their frequency in the general population, the local community surrounding a microbiology lab appears to be at no greater risk than the population at-large. For example, despite 109 lab-associated infections at the Center for Disease Control during the period 1947-1973, no secondary cases were reported in family members or community contacts. In sum, these data suggest that, while workers in microbiology labs are exposed to increased health hazards, the risk to the surrounding community is minimal.

As mentioned previously, applied genetics, especially recombinant DNA technology, has received more commercial promotion in the pharmaceutical industry than in other commercial sectors. For this reason, assessments to date of the potential risks arising from this new technology have been made in the context of laboratory and industrial practices pertinent to the pharmaceutical sector.

A number of risk assessments have been conducted attempting to evaluate the safety of using *E. coli* K12 as a host bacterium for the manufacture of human proteins via recombinant DNA techniques. Three conferences dealing with this issue have been held: (1) at Falmouth, Massachusetts, in June 1977; (2) at Ascot, England, in January 1978; and (3) at Pasadena, California, in April 1980. The viewpoints expressed at these sessions are summarized as follows:

> The natural fragility of K12 would make it very difficult, if not impossible, for it to colonize the human gut, or to be communicated between individuals.

> The transfer into K12 of genes encoding the manufacture of virulent proteins (toxins) would not produce a fully pathogenic K12 strain, and the insertion into K12 of DNA from human viruses would present fewer risks than the same viruses existing freely in nature.

> The ingestion of a K12 strain that synthesizes and secretes a human hormone, such as insulin, would not contribute significantly to the hormone levels that occur naturally. Even assuming that more efficient hormone-producing strains are developed, current procedures require that the secreted protein be attached to extraneous material that is removed during commercial processing of the product, but which would prevent the formation of an active hormonal substance in the gut of an individual. Future technical advances, however, may obviate this safety feature.

> The bacterial synthesis of human proteins in the GI tract (or elsewhere in the body) would not likely trigger an autoimmune response to the human substance.

That is, an individual infected with insulin-producing
K12 would not produce antibodies to human insulin.

Thus, the great bulk of evidence indicates that *E. coli* K12 is eminently safe as a host bacterium for mediating the synthesis of human proteins. There currently exists no firm evidence conflicting with this conclusion. Although one can speculate as to the risks arising from the concurrence of a variety of unlikely events, the experience accumulated so far indicates strongly that the risks are minimal or nonexistent.

Other microorganisms will soon be utilized as hosts for recombinant DNA procedures leading to the commercial production of drugs and biologicals. The two microbes most often discussed in this regard are a common soil bacterium, *Bacillus subtilis,* and brewer's yeast, *Saccharomyces cerevisiae.* As with *E. coli,* more is known of the genetics and molecular biology of these organisms than of any other microbes. Apprehensions regarding risks inherent in the use of these microbes have been far less than for *E. coli* K12. Neither *B. subtilis* nor *S. cerevisiae* cause serious infections in humans; only easily treated minor eye infections are attributed occasionally to *B. subtilis.*

Thus, these three microorganisms will likely underlie most commercial recombinant DNA activities within the pharmaceutical industry for the foreseeable future. Each of the three microbes has certain technical advantages and disadvantages that recommend its use on a commercial scale. The choice of which organism to use in a particular application will be made largely on economic grounds.

The NIH has recently approved the use of various species of *Streptomyces* as host organisms in recombinant DNA experiments (see 45 *FR* 50531). These microorganisms are especially important in the drug industry owing to their ability to manufacture the aminoglycoside class of antibiotics, including streptomycin, erythromycin, and tetracycline. The application of recombinant DNA technology to these microbial strains promises to generate improvements in product yield and, perhaps, to new and useful types of antibiotics.

NIOSH and NIH have examined the issue of worker safety in the pharmaceutical industry within the context of recombinant DNA activities. The NIH has proposed recommendations for large-scale fermentation of recombinant DNA organisms (analogous to the P1 to P4 designations for laboratory experimentation). Commercial firms are expected to comply voluntarily with these recommendations. So far, two U.S. firms, Eli Lilly and Genentech, have been granted NIH approval to proceed with scale-up operations.

In the spring of 1980, a NIOSH team conducted walk-through surveys of both these facilities. Eli Lilly operates a state-of-the-art commercial fermentation plant. All operations are closely monitored for leakage or contamination of biological material. Equipment is designed to minimize the formation of aerosols and to initiate sterilization procedures in the event of an accidental spill. Programs to ensure worker safety and health are in place, including medical surveillance, periodic safety inspections, monitoring employee work practices, and the provision of safety equipment and protective clothing.

Similar programs have yet to be instituted at Genentech, a firm that was founded in 1976 and has 100 years less experience than Lilly in large-scale fermentation operations. NIOSH, therefore, has recommended that Genentech plan immediately to implement similar safety and health protocols.

In summary, the pharmaceutical industry as a whole appears to be well equipped to deal with the various experimental and engineering safety issues that are posed by the advent of recombinant DNA technology. This industry historically has been involved in the "business of biology," and there exists a long tradition of safety associated with their operations. Moreover, a firmly established regulatory apparatus (largely housed in the FDA) already exists that closely monitors activities and screens new products originating from this industry. One must conclude that new products and processes stemming from various applications of genetic engineering in the pharmaceutical field will encounter the same careful scrutiny that has been devoted to conventional activities.

Chemical Industry

The near-term role of applied genetics in the chemical industry predicts that bioprocesses will be developed that perform chemical transformations on specific feedstocks to manufacture specialty products. Consequently, the industry will be compelled to engage in large-scale microbial fermentations in order to obtain the necessary reagents (either the organisms themselves or the enzymes they synthesize) to perform these chemical reactions. Such fermentations, and subsequent product isolation procedures, will proceed in a manner entirely analogous to similar operations in the pharmaceutical industry.

There exist differences, however, that may be of concern from an environmental or safety and health standpoint:

> The species of microorganisms likely to be utilized in the chemical industry differ from those in the drug industry. For example, various species of *Pseudomonas, Acinetobacter,* and *Flavobacteria* may find application in mediating chemical processes because these organisms naturally possess enzyme systems capable of catalyzing chemical reactions involving organic substrates

(such as petroleum products) that are of interest to the chemical industry. Many of these microbes are opportunistic pathogens in man; that is, they infest skin lesions or cause severe infections in individuals who are already weakened by a pre-existing ailment.

The chemical industry is unaccustomed to the application of biological processes as a business enterprise. Commercial-scale fermentations are alien to this industry. Chemical firms interested in adopting one or another bioprocesses may choose to purchase the technology, or to obtain the service through outside contract, rather than develop in-house capabilities.

The chemical industry has a poorer record than the pharmaceutical sector in areas related to worker safety and environmental protection. This discrepancy may reflect the grossly different commercial operations performed by these two industries rather than neglect. Nevertheless, one might be apprehensive of the introduction of a new technology into an industry where, historically, hazards have surfaced only after serious harm was done to workers or the environment.

In the long run, the replacement of conventional chemical processing steps with biological processes should serve to reduce the level of overall risks. The microbes or enzymes that mediate the bioprocess will be susceptible to inactivation by high concentrations of many organic feedstocks. Thus, feed streams will have to be diluted with nontoxic substances to obtain concentrations that permit survival of the biological systems involved. As a consequence of this dilution, the feed streams will become less toxic to workers who run the processes and to an environment that may encounter the stream in the event of a spill.

All currently envisioned applications of biotechnology in the chemical industry anticipate the use of closed bioreactor systems for performing discrete chemical reactions or for growing large volumes of microorganisms for use as biocatalysts or as sources of substitute feedstocks. Experience accumulated in the pharmaceutical sector indicates that routine operation of such systems poses minimal environmental hazard. A typical fermentation operation is depicted in Figure 3.1. Each step in the process, including double-sealed stirring rotors, positive pressure inside the vessel with loss-of-pressure alarms to warn of a breach in containment, and presterilization of all added materials, including air, antifoaming agents, acid, and base for pH control, is conducted to ensure sterility and containment. Furthermore, since the air vented from the fermentor generates aerosols containing microorganisms, the exit gas should also be sterilized. Although not performed routinely, this can be accomplished by passing the air through high-efficiency particulate air (HEPA) filters, or by exposing the gas stream to radiation, electrical discharge, or germicidal sprays.

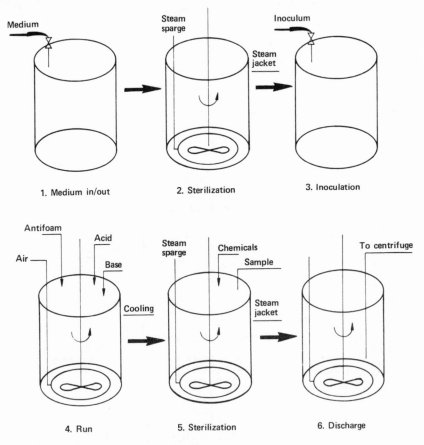

Figure 3.1: Steps in a typical fermentation process.

The fermentation process shown in Figure 3.1 involves sterilization of the reactor contents prior to sample work-up; that is, the microbes are killed before they are discharged from the vessel. The chemical industry might employ such a procedure in order to isolate an enzyme that the microbes have accumulated intracellularly or excreted into the medium. Figure 3.2 diagrams a process flow, including feed streams and waste streams, for isolation of an intracellular enzyme. The wastes from these processes consist of highly variable liquid streams containing high levels of suspended solids. These wastes typically have elevated chemical and biochemical oxygen demands (COD and BOD), as well

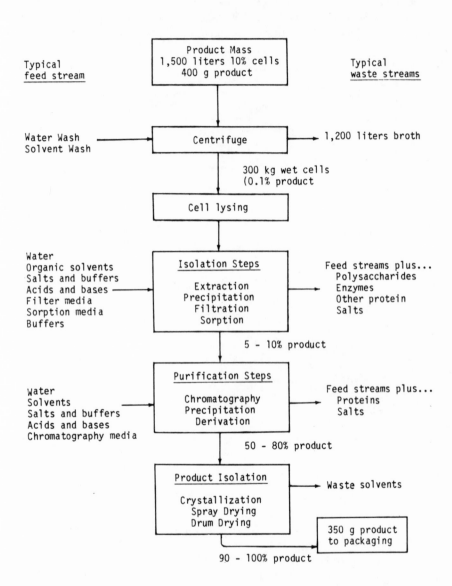

Figure 3.2: Product recovery from a typical batch fermentation.

as significant nitrogen and phosphate loadings. The pH is generally in the acceptable range, pH 5 to 9.

The application of biological processes in the chemical industry is in a very early developmental state. The ability of microorganisms, or their products, to mediate chemical transformations of organic substrates on a commercial scale has yet to be demonstrated. It seems probable that processes based on microbial systems will rely on activities that occur naturally among populations of microorganisms.

Energy Industry

The application of biological processes to the energy industry is at a very early stage of development. Other than the fermentation of ethanol from cornstarch for use in gasohol production, no commercial-scale bioprocess will have an impact on the energy sector for at least five years. The production of biofuels (e.g., ethanol, methane, vegetable hydrocarbons) from unconventional feedstocks has progressed only to the pilot scale, whereas biological hydrogen production is little more than a laboratory curiosity at present. Likewise, field tests have so far failed to demonstrate the general feasibility of using microbial systems for enhanced oil recovery. Thus, potential environmental hazards resulting from the use of applied genetics in energy production are highly speculative. Nevertheless, several comments are appropriate and some areas of potential concern can be identified.

> The production in the United States of sufficient ethanol to have a significant impact on domestic fuel supplies will require the diversion of enormous quantities of food crops, particularly corn. According to one estimate, a 4 billion gallon-per-year ethanol program could result in a 10 to 20 percent shortfall in corn supplies by 1990. This would severely limit the availability of grain for livestock feed or exports, thereby driving up food prices. Four billion gallons represent less than 5 percent of current annual fuel consumption. Clearly, useful alternative biomass feed-stocks for ethanol production are sorely needed.

> The prospect that cellulosic materials may serve as suitable feedstocks for biofuel production forewarns of large-scale deforestation, particularly in areas lacking alternative sources of biomass. Huge tracts of prime forest land in certain parts of the world have already been cleared for purposes of agriculture or fuel use. The wholesale conversion of wood biomass into ethanol threatens to exacerbate this trend.

Processes designed to convert lignocellulosic materials into substrates suitable for ethanol fermentation entail an initial hydrolysis step (see Figure 4.4). Hydrolysis can be accomplished either chemically, using strong mineral acids, or biologically with enzymes. The latter approach is preferable from a safety and environmental point of view but is less likely to be implemented in the near term. Thus, commercial processes generating large quantities of acid wastes can be anticipated.

As mentioned previously, the utilization of wastes and municipal sewage as raw materials for biogas generation promises to lessen the environmental burden imposed by these pollutants. Hazards may arise, however. If large centralized biogas facilities are planned, then one faces risks associated with the transport of the raw wastes to the site from various points of origin. A program to establish numerous local biogas generating stations may encounter variations in operating characteristics or in the level of personnel training that could mitigate against long-term safe operation of any particular facility.

The species of microorganisms likely to be utilized in enhanced oil recovery schemes–*Pseudomonas* and *Acinetobacter,* for example–are the same as those mentioned previously in the context of biotransformations of organic substances in the chemical industry. As already discussed, these microbes are potentially serious pathogens in man.

It is probable that the near-term use of microorganisms to mediate bioprocesses in the energy industry will exploit naturally occurring microbes. Thus, as is true for the chemical industry, the impact of genetic engineering (especially recombinant DNA techniques) will be minimal.

Finally, established energy companies are not accustomed to dealing with biological systems as a means of producing energy. (Most oil companies, however, do maintain some expertise in microbiology to assist in prospecting.) These firms will be compelled to strengthen their technical competence in areas related to biology as commercial prospects for bioenergy brighten. Hopefully, they will devote adequate attention to environmental hazards that may emerge from these new areas of business.

Mineral Industry

The limited scope of biotechnology in the mining industry confines the range of environmental concerns that demand consideration. However, all foreseeable applications of biological processes in this industry involve microbial systems operating in relatively open environments, such as slag heaps or tailings ponds. Consequently, there are risks that microorganisms or their metabolic products will inadvertently contaminate the local ecology. Specific areas of concern include the following:

> Bacterial leaching operations generate large quantities of sulfuric acid which, if poorly contained, could seriously contribute to the acidification of U.S. fresh water supplies.

> *Thiobacilli* and related bacterial species are not known to be pathogenic in man or animals; indeed, their peculiar metabolic characteristics suggest that they should be quite innocuous from a public health standpoint. However, increased use of genetically altered species on an industrial scale (and thus greater exposure to human populations) may select for bacterial strains that have acquired the ability to infect humans.

> The use of bacteria to concentrate metals from dilute waste streams of settling pond entails the risk that metals will accumulate in the food chain. Even though metal ions, such as mercury and silver, are highly toxic to bacteria, it is through microbial action that mercury, for example, is transformed into organic compounds that are responsible for mercury toxicity in higher forms of life. In other words, metals released into the environment are metabolized by naturally occurring bacteria. The key for safe commercialization of this bioprocess will be adequate containment of the operation to prevent dissemination of toxic metals into the general environment.

> As with the chemical industry, the mining industry has very little experience with biological processes. This lack of familiarity could result in a failure to recognize impending environmental hazards or in an eagerness to carry out biological processes before their safety has been firmly established.

Pollution Control Industry

Biological processes are currently in wide use throughout the pollution control industry, but so far modern applied genetics or genetic engineering has had negligible impact. The potential exists, however, that these new biotechnologies will drastically alter or replace conventional physical/chemical waste management processes. Nevertheless, considerable basic information regarding the relevant biological systems must be acquired before genetic engineering can be implemented to improve on naturally occurring organisms. Very little is now known of the biochemistry, metabolism, genetics, or natural ecology of the microbial species that mediate biodegradative processes. Indeed, the mere identification of potentially useful microorganisms is far from complete. Thus, the impact of applied genetics may not be felt in this industry for five years or more.

Nevertheless, increasing utilization of natural bioprocesses in pollution control entails certain potential hazards that are noteworthy and that may forewarn of future risks evolving from the application of genetic engineering in this industry.

Chief among these concerns is the generation of biological aerosols. These are tiny droplets of water or dust particles containing active microbial material. They remain suspended in the atmosphere to be transported by air currents to distances of several miles from their origin. Many industrial processes have the potential to create hazardous aerosols that contain pathogenic microorganisms. Among these are:

> Agricultural practices. Stockyards and poultry feedlots generate contaminated dust aerosols that may elicit very serious health problems, such as anthrax. The increasingly common practice of applying partially treated or untreated municipal sewage to crop lands has led to improved crop yields and has provided an alternative to the direct discharge of sewage into lakes and streams. But this practice gives rise to potentially harmful aerosols and to increased risk of ground water contamination. Waste water from food processing plants has also been utilized in land application programs.

> Textile mills. The processing of wool and animal hair produces dust aerosols that are known to contain pathogens, such as the causative agents of anthrax and Q fever.

> Abattoirs and rendering plants. The slaughtering and processing of livestock is a serious source of infectious aerosols, occasionally causing epidemics among employees. The condition is exacerbated by livestock farmers or ranchers who frequently rush their stock to market at the first sign of disease among members of the herd.

Sewage treatment plants are probably the most numerous and varied sources of pathogenic aerosols. The bubbling of air through an activated sludge facility and the splashing of sewage water over the rock bed of a trickling filter operation both generate numerous aerosolized particles. Approximately one-half of these droplets are in the size range (1 to 5 microns) that are carried downwind for considerable distances and which are readily inhaled and deposited in the human lung. The magnitude of the potential hazard posed by sewage treatment plants is related to the abundance of these facilities, their proximity to residential areas, the great variety of microbial species found in sewage, and the high frequency of aerosolization from these facilities which are in operation all day throughout the year.

Thus, the threat to public health posed by infectious aerosols is considerable. Moreover, many laboratory-associated infections also appear to result from the production of aerosols, rather than from more obvious lab incidents, such as pipetting by mouth, needle and syringe accidents, or simple spills. Clearly, future efforts to minimize risks associated with any microbiological process, including those involving recombinant DNA organisms, should focus on methods of controlling aerosols.

The biodegradation of organic pollutants by indigenous microorganisms is chiefly responsible for the eventual recycling of most environmental wastes. Organic pollutants fall into three general categories:

Completely biodegradable, for which there exist microbes capable of mineralizing the substrate. Examples include relatively simple hydrocarbons and aromatics, such as phenols.

Totally recalcitrant, for which there exist no known microorganisms capable of chemically transforming the substrate, or if so, at a very slow rate. Synthetic plastics, such as polyethylene or polyvinyl chloride, appear to be in this category, as do some polychlorinated aromatic hydrocarbons and pesticides.

Co-metabolized compounds are transformed to some extent by microbes that utilize other substances as sources of energy and biomass. Pollutants in this category, such as DDT, aldrin and heptachlor, are degraded slowly. Microbes have yet to be isolated that can use compounds of this type as nutrients.

The impact of applied genetics in this area will be minimal until more is learned of the types of microorganisms involved. The suitability of naturally occurring microbes for waste cleanup will be examined initially. Particular attention will be paid to *in situ* decontamination processes. However, several factors mitigate against widespread success in this area.

> The concentration of toxic chemicals at the site of a spill or dump site is often too high to permit survival of any microbe capable of degrading the pollutant.

> Concentrations of toxic chemicals sufficiently low to permit survival (generally less than 1,000 ppm) may be too low to sustain bacterial growth. Thus, additional nutrients must be supplied.

> Some toxic compounds, particularly those that are co-metabolized, are partially degraded into substances that are slightly less toxic than the parent compound but which are more readily mobilized (that is, dispersed throughout the local food chain). This is the fate of most polychlorinated organics.

Finally, there exist natural microbial systems capable of concentrating inorganic metal ions from dilute waste streams. This accumulation appears to be associated with chemical transformations of the metal into organic forms. The methylations of mercury and arsenic, for example, are known to occur as a result of microbial action in aquatic sediments. These organic derivatives are more toxic than the corresponding inorganic substances, and they are more readily taken up from the sediments by aquatic animals. Thus, commercial use of microbial systems to remove heavy metal ions from waste waters must be monitored for the release of even more toxic organic derivatives.

Agribusiness

Tables 3.2 and 3.3 list the potential environmental benefits and hazards that might be derived by applying genetic engineering technology to agricultural problems. The benefits and hazards are roughly listed in their order of importance with the most important being listed first. Secondary effects are listed towards the bottom of the list.

As can be seen, the positive aspects are already in the majority, and may produce benefits of enormous magnitude. Crop yield, as noted above, is crucial in regard to global food requirements and fertilization utilization is central to agricultural production, and pollution problems. The potential hazards although mostly speculative, cannot be easily dismissed. For example, decrease in crop variability has already produced a near disaster in corn production.

Table 3.2: Potential Environmental Benefits of Agricultural Genetic Engineering

Decrease nitrogen fertilizer input by conferring nitrogen-fixing ability on nonlegumes or by increasing the nitrogen-assimilating ability of nitrogen fixing microbes

Increase the amount of land under cultivation and increase marginal land usage

Increase tolerance of plants to environmental stresses (e.g., salts, pests, drought, frost, pollutants)

Enhance development of underexploited crops for multiple end uses (feed, fuel, fiber)

Improve yields in agriculture and silviculture by hastening cultivar selection and development

Increase crop uniformity and decrease culling of grains and vegetables

Improve nutritional value of seeds, fruits, and vegetables

Decrease cost of crop production

Decrease dependence on foreign supplies of oil and make farming less energy intensive

Enhance competitiveness of pioneer plant and microbial species (revegetation)

Decrease pollution and eutrophication of aquatic environments caused by pesticide and fertilizer run-off

Create new agricultural products (e.g., encapsulated pesticide-degrading microbes)

Decrease use of chemical pesticides by construction of effective microbial insecticides

Expand basic knowledge of gene structure, function, and regulation and the molecular mechanisms that control plant productivity

Table 3.3: Potential Environmental Hazards of Agricultural Genetic Engineering

Produce more vigorous weeds, and thus increase development and use of herbicides

Alter the niches and pathogenicities of plant viruses and soil bacteria

Increase breakdown of fertilizer and increase production of nitrogen oxides (denitrification)

Increase rate of organic matter decomposition (e.g., by *Erwinia*) in the soil

Decrease the range and variability of some crops, thereby inadvertently increasing disease susceptibility

REGULATION

Regulatory Background

Any precedent for regulating the genetic engineering industry will probably be based on the National Institutes of Health (NIH) guidelines for recombinant DNA research which are detailed in the *Federal Register* (NIH, July 7, 1981). The guidelines were first published in June, 1976, and are periodically updated. Additions and amendments are reviewed and recommended by the NIH Recombinant Advisory Committee (RAC) which consists of nongovernment members with expertise in scientific fields relevant to recombinant DNA technology and biological safety. In addition, the U.S. Department of Agriculture (USDA) has an ad hoc recombinant DNA committee, which submits its recommendations to NIH. In general, current containment levels are considered too restrictive for most plant research, and most recommendations have been for lower containment levels.

All institutions engaged in recombinant DNA research must establish an Institutional Biosafety Committee (IBC) to review projects for compliance with the NIH guidelines and for adoption of emergency plans covering accidental spills and contamination of personnel. Investigators are required to notify their IBC prior to initiation of experiments. The IBCs are responsible for reviewing the registration of documents for consistency with provisions of the guidelines and for advising the investigators on necessary changes. IBCs and investigators are still required to petition NIH's Office of Recombinant DNA Research directly about experiments that are not explicitly covered by the guidelines, that require case-by-case review, or that involve exceptions to the provisions.

NIH Physical and Biological Containment Requirements

Two types of containment are used in recombinant DNA research: (1) physical containment, which involves special procedures, equipment, and laboratory installations that provide physical barriers, and (2) biological containment, which limits either the infectivity of a vector or vehicle (plasmid or virus) for specific hosts or its dissemination and survival in the environment. Different levels of containment, appropriate for experiments with different recombinants, can be established by applying various combinations of physical and biological barriers, along with the constant use of standard procedures. The various physical and biological containment levels are described briefly in Table 3.4.

There are specific containment requirements for agricultural genetic engineering experiments. In shotgun experiments, genes from a num-

Table 3.4: Descriptions of Physical and Biological Containment Requirements

Category or Level of Containment	Description
	Physical Containment
P1	No special containment or laboratory design is required. All biological wastes shall be decontaminated.
P2	Biological safety cabinets shall be used to contain aerosol-producing equipment.
	Laboratory admittance is restricted.
	All biological wastes shall be steam sterilized before disposal.
P3	Biological safety cabinets shall be used to contain aerosol-producing equipment.
	Laboratory admittance is restricted and through a controlled access area.
	Controlled air movement in the laboratory is required.
	All biological wastes shall be steam sterilized before disposal.
P4	Experiments must be conducted in Class III cabinets or in Class I or II cabinets in which personnel are required to wear one-piece positive-pressure isolation suits. The area should be under negative pressure and the exhaust air filtered.
	Laboratory admittance is restricted and is through a clothing exchange and shower area. Laboratory clothing must be completely exchanged on each exit.
	The laboratory must have an individual air supply and exhaust system. Exhaust air will be HEPA filtered.
	An individual vacuum system with in-line HEPA filters is required. Water supply, liquid, and gaseous service shall be protected by backflow-prevention devices.
	All waste must be steam sterilized or decontaminated by suitable means before removal from the P4 facility.
	Biological Containment
HV1	Moderate level of containment. Includes *E. coli* K-12 or other prokaryotes comparable in containment.
HV2	High level of containment. Chances of the escape of recombinant DNA either via survival of recombinant organisms or via transmission of recombinant DNA to other organisms should be less than $1/10^8$.
HV3	Highest level of containment. All HV2 criteria are met. In addition, containment must be independently confirmed by tests in animals, and the relevant genotypic and phenotypic traits independently confirmed.

ber of free-living bacteria including *Azotobacter, Aspirillum, Rhodospirillum, Alcaligenes, Pseudomonas,* and *Escherichia* and from the symbiotic bacterium, *Rhizobium,* can be cloned and inserted into higher plant cells under relatively low containment requirements (HV1, P2, HV2 P1) (Table 3.5). The formation of DNA recombinants from cellular DNAs that have been purified can be carried out under lower containment conditions because there is little possibility that a modified virulent organism could escape from the laboratory. All the above organisms are Class I (low- or no-hazard) organisms (Office of Biosafety, HEW, 1974).

Containment conditions are higher for *Klebsiella, Clostridium* and *Salmonella* (HV2 P3) because they are known pathogens. Several species of *Clostridium* and all species and serotypes of *Klebsiella* and *Salmonella* are Class II (low-hazard) organisms (Office of Biosafety, HEW, 1974). *Clostridium* is commonly found in the intestinal tracts of man and other animals. Two species are known to cause botulism and tetanus. *Klebsiella* inhabits the human intestinal, respiratory, and urogenital tracts. One species causes pneumonia. All known forms of *Salmonella* are pathogenic to man and/or other animals, causing diseases such as typhoid fever, gastroenteritis, and enteric fever. One species of *Pseudomonas, P. pseudomallei,* is a Class III (moderate-hazard) organism (Office of Biosafety, HEW, 1974), that causes melioidosis or fatal blood poisoning. About ninety species of *Pseudomonas* are plant pathogens that cause leaf spot, leaf stripe, wilt, and similar diseases. Containment requirements for *P. pseudomallei* are EK1 P3, while requirements for the other species are HV1 P2. A proposal allowing DNA recombination experiments with Class III organisms and certain plant pathogens has been accepted (NIH, July 29, 1980). Formation of DNA recombinants using DNA from *Klebsiella, Clostridium, Salmonella,* or *Pseudomonas* that has been purified and in which the absence of harmful sequences has been established can be carried out under containment conditions lower than P3. Experiments with Class IV and Class V organisms are prohibited.

Genes from higher plant cells and plant viruses can be cloned under low containment requirements (HV1 P2 or HV2 P1). If a plant source makes a potent polypeptide toxin, containment is raised to HV2 P3. If a nonpolypeptide toxin is made, HV1 P3 or HV2 P2 is necessary. In shotgun experiments, genes for the toxin might be transmitted by mistake. A polypeptide toxin would be controlled by a relatively simple DNA sequence (probably one gene encoding the toxin). A nonpolypeptide toxin, on the other hand, would be controlled by a more complex set of DNA sequences and would probably not be the product of one gene. As previously noted, if purified DNA is used and the absence of harmful sequences has been established, lower containment conditions can be employed.

Table 3.5: Containment Requirements for Various Genes Using Certain HV1 and HV2 Systems*

Genes	Source	Containment**	NIH Guideline Section
Nitrogen fixation (nif)	*Rhizobium*	HV1 P2	III-A-1-b
	Klebsiella	HV2 P3***	III-A-1-b
	Azotobacter	HV1 P2	III-A-1-b
	Clostridium	HV2 P3†	III-A-1-b
	Aspirillum	HV1 P2	III-A-1-b
	Rhodospirillum	HV1 P2	III-A-1-b
Hydrogen uptake (hup)	*Alcaligenes*	HV1 P2	III-A-1-b
	Rhizobium	HV1 P2	III-A-1-b
Photosynthesis (cfx)	*Alcaligenes*	HV1 P2	III-A-1-b
	Higher plants	HV1 P2 or HV2 P1††	III-A-2-b
Water-splitting (lit)	Blue-green algae	HV1 P2†††	III-A-1-b
Biological stress (osm)	*Escherichia coli* K-12	EK1 P1	III-O
	Salmonella	HV2 P3***	II-A-1-b
	Higher plants	HV1 P2 or HV2 P1††	III-A-2-b
Cellulose utilization (cut)	Microorganisms	HV1 P2	III-A-1-b
Denitrification (den)	*Klebsiella*	HV2 P3***	III-A-1-b
	Pseudomonas	HV1 P2***,§	III-A-1-b
Plant viruses		HV1 P1	III-A-2-a

*Source: NIH, July 7, 1981.

**The formation of DNA recombinants from cellular DNAs that have been purified and in which the absence of harmful sequences has been established can be carried out under lower containment conditions than those used for the corresponding shotgun experiment (III-O).

***If DNA segments are entirely from *Klebsiella*, *Salmonella*, or *Pseudomonas* and propagation is within these organisms only, experiments are exempt from these guidelines (Appendix A of NIH Guidelines).

†HV2 P3 is required for *Clostridium botulinum*, *C. chauvoei*, *C. novyi*, *C. haemolyticum*, *C. histolyticum*, *C. septicum*, and *C. tetani*. All others require only HV1 P2.

††If a plant source makes a potent polypeptide toxin, containment is raised to HV2 P3. If a nonpolypeptide toxin is made, HV1 P3 or HV2 P2 is necessary.

†††Cloned segments of *Anabaena* DNA may be transferred into *Klebsiella* under P2 physical conditions (Appendix E of NIH Guidelines).

§EK1 P3 is required for *P. pseudomallei*. Others require only HV1 P2.

Containment for Vectors: Containment for the use of *Agrobacterium* as a vector is relatively high (P3) (Table 3.4) because it is a highly virulent plant pathogen that produces crown gall disease. *Agrobacterium* carrying new genes might have wide-ranging environmental consequences. Recently, however, some *Agrobacterium* experiments have been permitted at lower containment levels (NIH, March 12, 1981: November 21, 1980). While *Agrobacterium* is a virulent plant pathogen, plant tissue must be injured to permit infection. The genetic information introduced into plant cells by the *Agrobacterium tumefaciens* Ti plasmid is not transmitted by meiosis into seeds. DNA from plants and nonpathogenic prokaryotes may be cloned under P2 containment conditions. Transfer into plant parts or cells is permitted at the same containment level used for cloning in *Agrobacterium*. Containment levels have been reduced to P1 for well characterized genes, such as those that encode yeast alcohol dehydrogenase and maize zein.

Plant virus vectors, such as cauliflower mosaic virus and bean golden mosaic virus, have narrow host ranges and are relatively difficult to transmit mechanically to plants. Since they are unlikely to be accidentally transmitted by spillage of purified virus preparations, containment requirements are low (P1) (Table 3.6). Cauliflower mosaic virus is spread in nature by aphids, in which it survives only a few hours, while bean golden mosaic virus is spread by whiteflies. Neither is transmissible by seed or pollen.

Organelles, plasmids, and chromosomal DNAs used as vectors have a slightly higher containment requirement (P2 or P1) (Table 3.6) because there is a small possibility that DNA transmitted by such vectors could escape through pollen, seeds, or other propagules.

E. coli K12 has a low containment requirement (P1) (Table 3.6) because it is thoroughly adapted for use in the laboratory and has only a small chance of survival in nature. A proposal to include *Saccharomyces cerevisiae* for P1 containment was recently passed (NIH, July 29, 1980) since it also has only a small chance of survival in nature. However, containment should be modified to prevent spread of pollen, seeds, or other propagules if this is a possible escape route.

Rhizobium has not yet been proposed for use as a vector in any DNA recombination experiments. *Rhizobium* probably will require only a P1 containment. The host range of *Rhizobium* is limited to members of the Leguminosae.

Containment for Hosts: Cells in culture or plants growing in axenic culture as the host species generally require lower containment than intact plants because of the fragility of such hosts and the low survival rate of cultured cells outside the laboratory.

Generally, P1 or P2 containment is required for intact plant hosts. P1 conditions include a limited-access greenhouse or growth chamber that is insect restrictive, preferably made so with positive air pressure and with an insect-fumigation regime. In systems where propagules are used, negative air pressure probably is more appropriate. Soil, plant pots, and unwanted infected materials should be removed from the

Table 3.6: Containment Requirements for Various Types of Host Vectors*

| | .... Host. .... | | |
Vectors	Intact Plant	Tissue Culture Plants in Axenic Culture	NIH Guideline Section
Agrobacterium	P3**	P3**	Appendix E
Plant Viruses			
Cauliflower Mosaic Virus	P1	no containment***	III-C-3
Bean Golden Mosaic Virus			
Organelle, plasmid, or			
chromosomal DNAs†	P2 (P1)††	P1	III-C-4
Escherichia coli K-12	P1†††	P1	III-C-6
Saccharomyces cerevisiae	P1†††	P1	§
Rhizobium	P1	P1	§§

*Source: NIH, November 21, 1980.

**Cloned DNA fragments may be transferred to *A. tumefaciens* using a non-conjugative *E. coli* plasmid vector and into plant parts or cells in culture under containment conditions one step higher than required for the desired DNA in HVI systems (Appendix E). However DNA from plants and nonpathogenic prokaryotes may be cloned and transferred into plant parts or cells in culture under P2 containment. NIH recently granted permission to Mary Dell-Chilton, Washington University, to use *Agrobacterium* to transfer two well characterized genes (yeast alcohol dehydrogenase and maize zein) to tobacco plants using P1 containment (National Institute of Health, March 12, 1981). A proposal by Donald J. Merlo, University of Missouri, to introduce genes cloned in *E. coli* K-12 into *Arabidopsis* plants using *Agrobacterium tumefaciens* carrying an *E. coli*/Ti hybrid plasmid vector using P1 containment conditions was also recently accepted (NIH, November 21, 1980).

***Experiments using plant virus genomes to propagate DNA sequences from eukaryotic viruses require containment (See Table 3-5).

†Containment is required only when DNA recombinants are formed between vectors and DNA from cells other than host species. When DNA recombinants are formed between such vectors and host DNA and are propagated only in the host (or a closely related strain of the same species), experiments are exempt from the guidelines.

††P1 containment should be used if plants are too large for P2 facilities. But, run-off water should be sterilized if this is a plausible route for secondary infection, and standard P2 practices should be used for microbiological procedures.

†††Containment should be modified to prevent the spread of pollen, seeds, and other propagules. A request by Ron Davis, Stanford University, to field test corn plants into which corn recombinant DNA has been added using *E. coli* or *S. cerevisiae* vectors is still pending.

§ A recent proposal to include *S. cerevisiae* with *E. coli* for P1 containment was passed (NIH, July 20, 1980).

§§ At this time, no research has been proposed using *Rhizobium* as a vector, so it is not included in the Guidelines. However, NIH officials state that it would probably be assigned a P1 containment level (W. Gartland, personal communication).

greenhouse or cabinet, sealed in insect-proof containers, and sterilized before disposal. Run-off water should also be sterilized, if this is a potential escape route.

Proposed Regulations

On February 8, 1982, the RAC passed a proposal to lessen the administrative requirements of the guidelines. If finally approved by the NIH Director, the new guidelines would:

> (1) simplify section III to contain 4 classes of experiments (a) require RAC review before initiation (b) require only local Institutional Biosafety Committee (IBC) approval before initiation (c) require IBC notification only and (d) those experiments exempt from the guidelines;
>
> (2) lift 2 of the five current prohibitions (work with pathogens or cells infected by pathogens and large scale work with *E. coli, b. subtilus* and *s. cerevisiae*).

As with public perception of the possible benefits resulting from applied genetics, considerations of potential risks associated with these technologies have focused on recombinant DNA procedures. As mentioned, several workshops have been held during the past four years to review and summarize the status of risk assessment in the recombinant DNA field. In addition, the NIH (through its Recombinant DNA Advisory Committee, the RAC) has prepared a Risk Assessment Plan which will summarize and update annually information relevant to recombinant DNA risk assessment. The most recent update was published in the December 7, 1981 issue of *Federal Register*. A principal component of the plan is to analyze risk data pertaining to three general categories of host-vector systems in common use: prokaryotic (e.g., *E. coli*, K12), lower eukaryotic (e.g., *Saccharomyces cerevisiae*), and higher eukaryotic (e.g., mammalian cells). The following excerpt summarizes the current understanding of the risks:

> . .despite intensive study by the RAC Subcommittee on Risk Assessment and NIH staff, several conferences and workshops to consider specific issues and several experiments, no risks of recombinant DNA research have been identified that are not inherent in the microbiological and biochemical methodology used in such research (45 *FR* 61874).

Thus, in the absence of data of any compelling evidence to the contrary, and despite assiduous efforts to identify any potential hazards, scientists are now convinced that the practice of recombinant DNA techniques in the research laboratory setting poses no health risks over and above those encountered in normal microbiological research. However, wide array of industrial uses of applied genetics can be grouped into two categories with respect to environmental issues: (1) those applications in any industrial sector that may constitute an adverse impact on the environment; and (2) bioprocesses designed to assist in the effort to control pollution and constitute a net positive impact on the environment.

Since the NIH committee charged with the task of monitoring the field of recombinant DNA has reached the conclusion that there exist no untoward risks in practicing this technology, what will become of the committee itself? Indeed, what is the future of government involvement generally in this area? Several comments and recommendations can be made:

> The IBCs are naturally reluctant to take on the added workload in dealing with what are now deemed innocuous safety issues.

> As are other governmental agencies, the RAC is beginning to examine issues from a cost/benefit standpoint. The RAC recognizes that other areas of potential health concern exist in the biomedical research field. These more conventional hazards, which exceed the threat of recombinant DNA as potential risks, include exposure to pathogenic aerosols, X-rays, radionuclides, and toxic chemicals. Any future role for the RAC (or some other RAC-like organization), perhaps an IRAC (Industrial-RAC), should include consideration of these safety issues as a priority.

> Biotechnologies other than recombinant DNA have so far received little attention with regard to potential hazards. For example, cell fusions involving various cells derived from human tissues may become increasingly popular as a method for obtaining human biologics for drug manufacture. Large-scale application of these procedures entails the speculative risk that pathogenic viruses will be induced and propagated. Relevant government agencies should be advised to monitor the application of any biotechnology within their purview.

> The commercial applied genetics industry is at a nascent stage of development, and, so far, no incidents of environmental concern related to this industry have

materialized. Any environmental risks arising from industrial use of applied genetics are speculative. At this time there exists no compelling reason to establish regulations in this area.

Should environmental hazards emerge in the future, it is probable that they can be handled within the existing regulatory framework. A consensus appears to have been reached that the authority to regulate commercial uses of this technology exists, including: (1) requirements for premanufacture review of industrial processes based on recombinant DNA methodologies (Section 5 of TSCA); (2) restriction or prohibition of manufacture, processing, distribution, or use of recombinant DNA if such is deemed hazardous (Section 6); and (3) dealing with imminent hazards involving recombinant DNA (Section 7). Moreover, the discharge of recombinant DNA material into the environment could be regulated under existing statutes within the Clean Air and Water Acts. In summary, no additional legislation would seem to be necessary.

The Federal Government should continue to take an active role in promoting applied research and development of biological waste management processes and techniques. Emphasis should be placed on the biology of relevant systems rather than on process engineering and design. A particularly troublesome problem requiring more research is *in situ* decontamination of chemical wastes. A more tractable problem deserving attention involves the use of on-line bioreactors for treating industrial effluents at the source.

Further investigation is needed into the generation, dispersal, and control of biological aerosols.

To the best of its abilities the Federal Government should monitor commercial and scientific developments in the field of applied genetics with the aid of identifying both imminent environmental hazards and areas where this technology might be applied to pollution control operations.

4

Industrial Applications, Trends, Potential Hazards

This chapter contains an industry-by-industry analysis of biotechnology applications. Each industrial sector will be examined for:

Current activities in applied genetics. Some speculation may be required inasmuch as certain information is held as proprietary by private industries.

Future prospects for the application of biotechnology within each industry.

The following commercial sectors will be examined: (1) pharmaceuticals, (2) industrial chemicals, (3) energy, (4) mining, and (5) pollution control.

The Electronics Industry is beginning to show an interest in applications of Biotechnology to production of electronic components. Studies in progress are demonstrating the potential for thousand-fold multiplication of storage capacity on a silicon chip. However, at this time this work is only at a preliminary state and is being actively pursued by the Office of Naval Research and one small firm. (EMV Associates; Table 6.2.)

PHARMACEUTICAL INDUSTRY

Current Activities and Future Prospects

The largest efforts to date towards commercial application of modern biological techniques have taken place in the pharmaceutical industry. The manufacture of new or improved drugs and vaccines surely will be first commercial fall-out from recombinant DNA technology. Scientists recognized very early in the development of these techniques the immediate potential for mass-producing human biologicals, such as hormones and serum proteins, for eventual use as therapeutic agents. If available at all, such agents traditionally have been isolated from animal sources, a practice frequently leading to shortages in supply or to variation in quality from batch to batch. Moreover, biologics derived from animals generally differ slightly in structure from the human form of the analogous compound, thus providing a less than optimal human therapeutic.

The production by modern biotechnological methods of specific pharmaceuticals will now be considered.

Interferon is a protein synthesized by most cells of higher organisms in response to virus infections. Its production and secretion in miniscule amounts by infected cells serves to "interfere" with the spread of the infection to healthy cells. Thus, the administration of interferon as a drug promises to be a potent antiviral therapy. In addition, interferon has been shown to act as an antitumor agent for certain types of cancer. Its potential as a cancer drug is now under thorough investigation at several clinical centers in the United States, notably the M.D. Anderson Hospital and Tumor Institute in Houston, the Memorial Sloan-Kettering Cancer Center in New York, the Sidney Farber Cancer Institute in Boston, the Stanford University Medical Center, and the National Cancer Institute.

The severe shortage of purified human interferon has hampered adequate testing of its therapeutic value, but scientists have succeeded in applying recombinant DNA techniques to create bacterial interferon "factories" that promise to increase greatly the supply of the drug, while reducing enormously its current cost of several thousand dollars per dose. A predicted market of $3 billion per year has lured numerous commercial firms, both in the United States and overseas, to invest huge sums in interferon production and testing. Some companies are pursuing tissue culture methods, rather than recombinant DNA techniques, to obtain usable quantities of interferon. The therapeutic and commercial values of interferon will likely be revealed within the next year or two.

Preliminary clinical results of interferon testing on cancer patients have been equivocal. Interferon appears to be no better than available chemotherapeutic agents in controlling certain prevalent types of cancer, such as multiple myeloma, breast cancer, or melanoma. Moreover, interferon-treated patients have suffered some of the same adverse side effects, including hair loss and nausea, typically associated with most drugs currently used in cancer treatment. Some patients have developed immune responses to interferon—a finding which, if shown to occur frequently, could greatly curtail widespread use of interferon in cancer therapy.

Clinical studies are also underway to assess interferon's value as an antiviral agent for the potential treatment of numerous ailments ranging from the common cold to encephalitis. So far, the drug has shown considerable therapeutic effectiveness in these applications, but no greater than that shown by certain new antiviral agents, namely ribavarin and didemnin, both of which are relatively simple chemical compounds that are far less costly to produce than interferon.

Insulin, a hormone made in the pancreas, aids in the metabolism of sugar. It is composed of two small polypeptides, the A and B chains, which are composed of 21 and 30 amino acids, respectively. In the pancreas, proinsulin is made as a precursor to insulin. Prior to secretion, the proinsulin is converted to insulin by the enzymatic removal from the middle of the molecule of a stretch of 35 amino acids called the C chain.

There are currently several strategies for producing insulin by recombinant DNA technology. Here we discuss two of them. In one, the genes for the A and B chains are chemically synthesized separately and inserted into separate plasmids as fusion proteins joined to the lac operon enzyme, beta-galactosidase. The gene to be cloned is a combination of the gene for the A or B chain and the gene for the enzyme, joined by the triplet codon for the amino acid methionine. The plasmid is then cloned in a bacterial host. Since neither the A nor B chain contains methionine, it can be efficiently removed from the fusion protein after the protein is extracted from the host. Removal is accomplished by treating the fusion protein with cyanogen bromide, which cleaves at the methionine juncture. The A and B chains are bound together as insulin by two disulfide bonds. After extraction from the enzyme proteins, they can be joined in the laboratory by using an air oxidation technique involving S-sulfonated derivatives and an excess of A chain. This methodology is 50 to 80 percent efficient in making the complete insulin molecule.

The second method utilizes only one organism to produce a fusion protein containing proinsulin. As in the first method, the extraction is made with cyanogen bromide. The isolated molecule is then treated with enzymes to remove the C chain, and the active insulin is recovered.

Figure 4.1 shows schematically the synthesis of proinsulin and insulin by recombinant DNA methodology.

Human insulin produced by gene cloning methods is the first "recombinant" drug to undergo clinical testing on human patients. According to Eli Lilly & Co., the drug's manufacturer, the synthetic insulin is biologically active in several test systems and its chemical and physical structures are indistinguishable from natural human insulin. But Lilly is facing stiff competition in the insulin market from the Danish Novo Industri, who have succeeded in cheaply producing human insulin by modification of pork insulin. This drug is also undergoing clinical trial in Europe and the United States. However, both approaches to the treatment of diabetes may eventually be obviated by successful transplantation of insulin-producing cells from the pancreas of nondiabetic donors. This experimental procedure has been demonstrated in lab animals and application to humans may be feasible within five years.

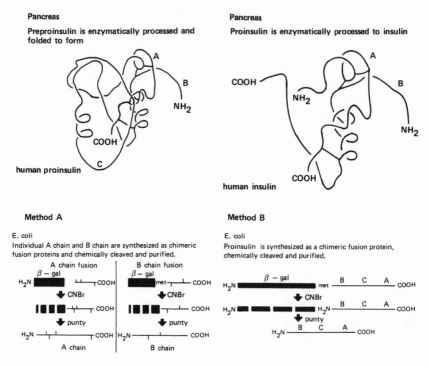

Figure 4.1: Alternative methods for insulin production in *E. coli.* (Source: Ross, M.J., 1980.)

Human growth hormone (hGH) or somatotropin is produced in the pituitary gland and mediates growth and stature, particularly in children. The hormone traditionally has been extracted from the pituitaries of human cadavers (animal substitutes are not suitable) and is used in the treatment of dwarfism in children. Recombinant DNA technology offers the prospect of sharply increased supplies of the scarce hormone, leading to speculation that hGH will be useful in the treatment of a variety of disorders including ulcers, burns, bone fractures, and bone deterioration (osteoporosis, a common ailment of the elderly). Moreover, hGH may stimulate growth in a group of children (numbering close to a million in the United States) who are abnormally small despite having seemingly normal levels of circulating growth hormone. Clinical trials of "recombinant hGH" by Genentech are well underway at ten U.S.

medical centers and, under license to Kabi A.G., preliminary tests are beginning in Sweden. Initial evidence that the drug induces growth in undersized children should be available by mid-1982.

Human growth hormone has been sequenced in its entirety. The synthesis of an expression plasmid for bacterial hGH synthesis involved cloning a synthetic DNA fragment coding for the first 24 amino acids separately from a clone coding for the remaining 167 amino acids. The nonconjugable plasmid pBR322, which codes for resistance to the antibiotics ampicillin and tetracycline, was used as vector for both clonings. The combined hybrid gene for the entire 191 amino acids was fused to the gene for beta-galactosidase in the lac operon and then inserted into a new expression plasmid subsequently designated pHGH107. Figure 4.2 shows schematically the stages involved in constructing the final expression plasmid coding for the complete amino acid sequence of hGH.

The synthetic DNA segment coding for the 24 amino acids was constructed from 16 chemically synthesized lengths of DNA. These fragments were joined together using the enzyme T4 ligase. The resulting 84-basepair fragment was designed to have sticky ends by adding additional nucleotides at each end and then treating it with the restriction endonuclease enzymes *Eco* RI and *Hind* III. *Eco* RI and *Hind* III were also used to open the vector, plasmid pBR322. The synthetic fragment was inserted into the plasmid and subsequently cloned. Then the plasmid with the correct DNA fragment was isolated from a cloned colony and designated pHGH3.

The 501-basepair cDNA fragment coding for the remaining 167 amino acids was prepared from pituitary mRNA, treated with the restriction endonuclease enzyme *Hae* III, and tailed with chemically synthesized segments of cytosine (C) nucleotides. The plasmid was treated with *Pst* I and joined to synthesized segments of guanine (G) residues. The vector and the fragment, rendered complementary under these conditions, were then joined together. Insertion and cloning followed.

In order to clone the complete gene, the two fragments were isolated from their plasmids and then joined together. The shorter, synthetic piece was cleaved from its plasmid with *Eco* RI and *Hind* III and then treated with *Hae* III to produce an *Eco* RI sticky end at one end of the fragment and a *Hae* III blunt end at the other. The larger cDNA fragment was then cleaved with *Hae* III and *Xma* I to produce a *Hae* III blunt end and an *Xma* I sticky end. The complete gene was made by joining the two *Hae* III blunt ends of the fragments with T4 ligase. Simultaneously, the *Xma* I end of the larger fragment was blunted with *Sma* I. This made that end suitable for insertion into a new plasmid (pGH6), as shown in the figure. This plasmid had been previously cloned with a copy of the lac operon. It was opened with *Eco* RI and *Hind* III and treated with SI nuclease, thus leaving the plasmid with one *Eco* RI sticky end and one blunt end. The complete hGH gene was then fused to the lac operon, which permitted the expression of the hGH gene in the presence of lactose.

1. Cloning of Synthetic Fragment

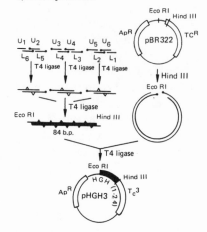

2. Cloning of cDNA Fragment

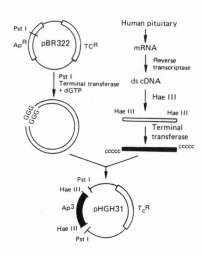

3. Assembly of HGH Gene

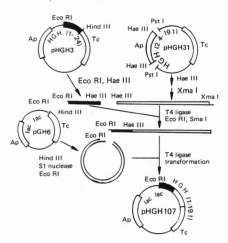

Figure 4.2: Construction of a bacterial plasmid coding for the synthesis of human growth hormone. (Source: Miozzari, G., 1980).

A number of other human peptides have been synthesized using recombinant DNA techniques. These include:

Somatostatin, a short fourteen amino acid peptide hormone secreted by the hypothalamus, was the first human substance produced in bacteria. It may have therapeutic potential in the treatment of diabetes.

Gonadotropin-releasing hormone, another hypothalamic hormone, is under investigation as a male contraceptive.

Vasopressin, also from the hypothalamus, appears to stimulate memory and learning in both healthy individuals and in certain mentally disturbed patients.

Thymosin, a thymus hormone, regulates the development of a portion of the immune system. As a potential drug, it may influence the aging process and have application in cancer therapy.

Beta-endorphin is a naturally occurring opiate that mimics the action of morphine. It has considerable therapeutic potential as a safe, nonnarcotic pain-killer.

Urokinase, a kidney enzyme, dissolves blood clots. It has potential as a drug to reduce the likelihood of heart attacks and strokes.

Serum albumin, the most prevalent protein in blood serum, is needed in large quantities to treat victims of shock and trauma.

Relaxin, a hormone that softens the cervix during pregnancy, may find use in easing childbirth. Relaxin may also have application in the treatment of rheumatoid arthritis.

Lymphokines and cytokines are potent bioactive peptides that regulate the activities of the immune system and that have been implicated in defense mechanisms against cancer (interferon is one example of a lymphokine).

A second major pharmaceutical area in which recombinant DNA techniques are finding considerable application is in the development of new vaccines. Conventional vaccinations against viral diseases involve immunizing with inactivated virus particles, which stimulates the host's immune system to defend against a subsequent exposure to a live, active virus infection. The use of entire viruses as the immunizing agent, however, entails the risks that either the vaccine may elicit the disease (owing to incomplete inactivation), or that the vaccine will be ineffective as a result of denaturation of the virus during inactivation.

Medical scientists have acquired an understanding of the molecular basis of vaccination, so it has become possible to isolate the specific proteins from the outer surface of viruses that are responsible for stimulating an immune response. Injection of these proteins alone is sufficient to generate adequate immunity to the viruses, but the proteins are totally nonpathogenic; that is, they do not mediate an infection. Using recombinant DNA techniques, it has been possible to clone the viral DNA that directs the synthesis of these proteins and to gain expression of the genes in bacteria so that the proteins are manufactured. Such research has focused on efforts to generate vaccines to immunize against:

> Hepatitis, a serious liver disease that affects an estimated 500 million people around the world.
>
> Influenza, the many forms of which have made reliable vaccines unobtainable using conventional techniques.
>
> Foot-and-mouth disease, a life-threatening disease among domesticated livestock. This joint project of Genentech and the U.S. Department of Agriculture will likely generate one of the first products of gene cloning technology to reach the marketplace. If effective, the vaccine should prevent annual losses measured in the billions of dollars worldwide.

In addition, vaccines are under development to combat certain pathogenic bacteria and the diseases that they cause, including:

> Gonococcus, which causes veneral disease
>
> Pathogenic *E. coli,* which give rise to digestive ailments such as severe diarrhea, of life-threatening concern in infant children
>
> Oral bacteria, which are responsible for tooth decay

The discussion of applied genetics in the pharmaceutical industry has so far centered on the uses of recombinant DNA technology. A variety of other biotechnologies are finding application in this industry as well, including:

> Monoclonal antibodies for use as diagnostic agents for viral and parasitic diseases, such as hepatitis and malaria, and for rapid identification of various allergic conditions
>
> New antibiotics generated by combining the synthetic capabilities of different antibiotic-producing strains
>
> Bacterial production of chemical intermediates for use in drug synthesis, such as glutathione, a liver drug intermediate

The growth of cancer patients' own tumor cells in tissue culture to permit safe testing of the most effective drugs to treat the specific tumor

The mass production of various important drug substances, such as interferons and blood clotting factors, by tissue culture of normal human cells

The production of human serum proteins by tissue culture of cells derived from fusions between human embyronic cells and mouse liver tumor cells

Fetal diagnosis of sickle cell anemia and other genetic diseases using techniques that distinguish highly similar DNA segments according to their patterns of fragmentation by restriction endonucleases

The production of vitamin B_{12} by bacteria, which should prove more economic than its isolation from fungi, as currently practiced

Chemical modifications by microbes of drug intermediates, as in the synthesis of various antibiotics (streptomycin, penicillin, and gentamycin) and the transformation of steroids towards the manufacture of contraceptives

The use of transposons to manipulate antibiotic-related genes in microorganisms to elicit overproduction of antibiotic intermediates. (Transposons are genetic elements that facilitate the movement of adjacent genes from one site on the DNA to another.)

The use of lipid vesicles (liposomes) as nontoxic, biodegradable and nonantigenic delivery systems for targeting drugs to their site of action (e.g., tumors)

The use of higher plants and sea creatures for the production of steroids, antibiotics, atropine, digitalis, etc.

A great variety of pharmacologically active agents can be isolated from naturally occurring microorganisms; a partial list is shown in Table 3.1.

CHEMICAL INDUSTRY

Current Activities

While not attracting public attention to the extent that interferon has, the chemical industry has been influenced by recent advances in biotechnology. Moreover, the industry may be on the verge of a technical revolution in which biological processes and renewable resources will rapidly replace the physical-chemical transformations of petroleum feedstocks upon which the industry is currently based. This section will attempt to outline some of the applications of biotechnology that are now in use and which serve as prototypes for the kinds of bioprocesses that may soon pervade this industry.

One chemical process utilizing biotechnology that has received some attention is under development by Cetus in conjunction with Chevron Oil. The process entails oxidation of alkenes to the corresponding alkene oxides. These end-products are utilized in enormous quantities for plastics manufacture; for instance, ethylene oxide and propylene oxide are the raw materials for the production of polyethylene and polypropylene, respectively. The Cetus/Chevron bioprocess consists of three enzyme-catalyzed steps, as follows:

$$\text{I.} \quad \text{Glucose} + O_2 + H_2 \xrightarrow{1} \text{fructose} + H_2O_2$$

$$\text{II.} \quad \text{Propylene} + H_2O_2 + \text{KBr} \xrightarrow{2} \text{bromoisopropanol}$$

$$\text{III.} \quad \text{Bromoisopropanol} \xrightarrow{3} \text{propylene oxide} + \text{KBr}$$

$$\text{enzyme 1} = \text{glucose oxidase}$$
$$\text{enzyme 2} = \text{chloroperoxidase}$$
$$\text{enzyme 3} = \text{halohydrin epoxidase}$$

The process is undergoing pilot plant scale-up. (In June of 1982 Chevron announced they were terminating their support of the fructose project with the Cetus Corporation.) The plan calls for designing an immobilized enzyme bioreactor in which the three enzymes are stably linked to an inert matrix. A continuous flow process ensues in which starting materials are percolated through the reactor, and products (fructose and alkene oxide) are recovered at the reactor outlet. It remains to be seen if this process is economically competitive with conventional alkene oxidations. Moreover, since the alkene starting material will generally be obtained from petroleum feedstocks, the process fails to overcome the dependence on dwindling and ever-more-costly oil supplies.

The general use of microbial enzymes in industrial processes (Table 4.1) is rapidly becoming a big business. One estimate places the 1985 market in enzyme technology at $500 million. The food industry historically has been the primary user of enzyme-based processes, and will continue in this role as demand increases for sweeteners derived from cornstarch and from other less conventional forms of biomass. But rising demand for gasohol will lead to further uses for enzymes in ethanol production. Three general classes of enzymes are finding increasing commercial use:

Amylases break down polysaccharides, such as cellulose, and mediate biomass conversions.

Proteases break down proteins and are used commonly in the food industry, for example as meat tenderizers.

A miscellaneous group, which includes oxidases and isomerases capable of performing specific chemical transformations of substrates, may soon find considerable utility in the chemical industry.

Table 4.1: Commercial Uses of Enzymes

Enzyme	Uses
Proteases:	
Alcalase	Detergent additive to remove protein strains
Bromelain	Meat tenderizer
Papain	Stabilize chill-proof beer; meat tenderizer
Pepsin	Digestive aid in precooked foods
Trypsin, ficin, and streptodornase	Wound debridement
Rennin	Cheesemaking
Carbohydrases:	
Amylase	Digestive aid in precooked foods
Amyloglucosidase	Production of dextrose from starch
Cellulase and hemicellulase	Preparation of liquid coffee concentrates and conversion of cellulose to sugar
Glucose isomerase	Production of high-fructose syrups
Invertase	Prevention of sugar granulation
Lactase	Prevention of lactose crystals in ice cream
Pectinase	Clarification of wine and fruit juices
Catalase	Peroxide removal in cheesemaking
Lipase	Flavor production in cheese
Lipoxygenase	Bread whitening

These industrial processes utilize microbes as sources of biological catalysts (enzymes) that, in turn, convert organic starting materials into products. A large variety of microorganisms directly synthesize simple organic chemicals when grown on carbohydrate substrates (see Table 4.2). Since many of these compounds are toxic at relatively low concentrations, considerable research effort is being expended to generate microorganisms that tolerate higher doses of these organics. Also, modern fermentation technologies, such as continuous flow and solid state processes, will be useful here since metabolic products never accumulate to poisonous levels. In addition, the use of unconventional substrates for microbial fermentation, such as cellulose and lignin wastes, is rapidly becoming feasible. These technical advances may soon make economic bioproduction of these and many other organic compounds possible.

Table 4.2: Organic Compounds Obtainable by Microbial Fermentation

Compound	Structure
Acetone	$CH_3\overset{\displaystyle O}{\overset{\displaystyle \|}{C}}CH_3$
Acetic acid	CH_3COOH
Acrylic acid	$CH_2{=}CHCOOH$
Butanol	$CH_3CH_2CH_2CH_2OH$
Citric acid	$HOOC-\underset{\displaystyle OH}{C}-(CH_2COOH)_2$
Ethanol	CH_3CH_2OH
Ethylene glycol	$HOCH_2CH_2OH$
Furfural	
Gluconic acid	$COOH$ $(CHOH)_4$ CH_2OH
Glycerol	$HOCH_2\underset{\displaystyle OH}{CH}CH_2OH$
Isopropanol	$CH_3\underset{\displaystyle OH}{CH}CH_3$
Itaconic acid	$COOH$ $C{=}CH_2$ CH_2COOH
Keto-gluconic acid	$COOH$ $C{=}O$ $(CHOH)_3$ CH_2OH

(continued)

Table 4.2: (continued)

Compound	Structure
Lactic acid	COOH | CHOH | CH_3
Malic acid	COOH | CHOH | CH_2COOH
Methanol	CH_3OH
Propionic acid	CH_3CH_2COOH
Tartaric acid	COOH | $(CHOH)_2$ | COOH

A number of microbial species growing on carbohydrate are able to synthesize surfactants or detergents. These compounds are typically long-chain fatty acids; current commercial production of surfactants requires petroleum feedstocks. The British sugar producer, Tate & Lyle, is now engaged in pilot-scale development of this process.

A variety of new industrial biopolymers are receiving attention for a wide range of uses: in the food industry as thickeners, in the petroleum industry for the secondary recovery of oil, and in various printing and coating applications. These materials generally are polysaccharides that are extractable from trees, seeds and seaweeds, or that can be manufactured by modifying starch or cellulose. But the most widely discussed biopolymer of this type is xanthan gum, an extracellular polysaccharide secreted by the bacterium, *Xanthomonas campestris*. Xanthan's unique gelling properties are useful in instant desserts and in salad dressings that flow readily out of the bottle but stick to the salad. Considerable enthusiasm greeted the prospects for xanthan gum to increase water viscosity in efforts to displace bound oil from rock formations in spent oil wells. However, field tests of the technique have so far failed to demonstrate the suitability of xanthan in this application.

Bacterial biopolymers are also under investigation for use as novel plastics. One such polymer is polyhydroxybutyrate (PHB) which is synthesized naturally by various bacterial species, including *Alcaligenes eutrophus, Azotobacter vinelandii,* and *Pseudomonas solanasearum.* PHB has physical properties similar to polypropylene, but its commercial production would be far more expensive than making comparable plastics from conventional petroleum feedstocks.

Photosynthetic algae offer the prospect of direct conversion of sunlight into useful organic chemicals. A considerable variety of end-products may be obtained from marine algae as shown in Figure 4.3.

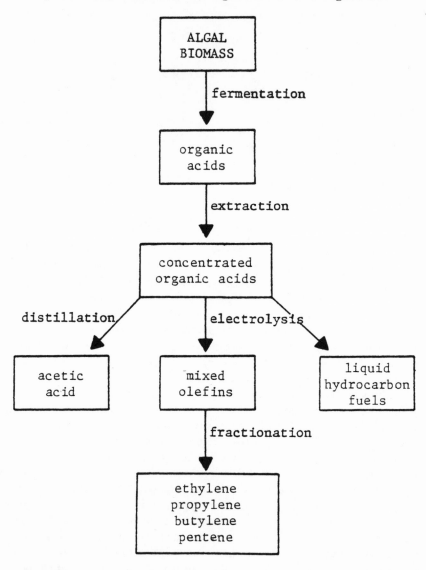

Figure 4.3: The extraction of useful chemicals from algae. (Source: Sanderson, J.E., et al., 1979.)

Whereas certain microorganisms can synthesize various simple organic compounds, some higher plants have acquired the ability to manufacture rather complex molecules. As shown in Table 4.3, these substances include rubber and petroleum substitutes, insecticides, steroids, and other drug precursors. The guayule is a scruffy bush native to the American Southwest and Mexico that makes a natural rubber capable of substituting for hevea rubber. The U.S. imports annually more than $1 billion worth of natural rubber from Southeast Asia. American tire company officials project that the cultivation of 100,000 acres with guayule (requiring an initial investment of more than $50 million) could produce 50 million pounds of rubber annually. Both the Goodyear and Firestone companies are investing modestly in this area, but without help from the U.S. government, a commercial guayule industry is 15 to 20 years away. Another desert shrub, the jojoba, propagates a bean whose oil is an excellent substitute for sperm whale oil in cosmetics, drugs, plastics, lubricants and waxes. Many large U.S. firms, including Pennzoil, Mobil, Westinghouse and Revlon, are investigating the many potential uses of jojoba oil, and several small start-up firms have emerged recently with the intent to commercialize the shrub. Euphorbia, a weed native to the U.S. Pacific coast, produces various hydrocarbon-like substances in large quantities that can be refined into fuels and feedstocks. The Diamond Shamrock Corp. has invested $2 million in a five-year program to examine the commercial potential of euphorbia, and a new start-up firm, the Genetic Oil Co. (Genoco), has established facilities in California to investigate the possibility of improving euphorbia through genetic engineering.

The potential applications of genetic engineering in the chemical industry lie largely in the area of organics production. Many organic chemical feedstocks can be produced utilizing fermentation technology (see Table 4.2). In principle, the efficiency of microbes in any fermentative process can be improved by using recombinant DNA techniques or other biotechnologies. The extent to which biological processes will supplant chemical processes in the chemical industry will surely be a function of economics; the future cost of petroleum will be particularly influential. Though the entire chemical industry uses only 7 percent of the petroleum supply in the United States, this industry is heavily dependent upon this resource.

A major segment of the chemical industry is engaged in the manufacture of organic polymers, including various plastics, fibers and synthetic rubbers. This industry is almost wholly dependent on petroleum feedstocks to provide the chemical monomers necessary for polymer production. However, microbial production of a few such monomers may eventually replace petroleum-based manufacture. Among the most likely polymers in this category are: polyamides (chemically related to proteins), acrylics, polyisoprene rubbers, and polystyrene.

Table 4.3: Examples of Useful Chemicals Derived
from Plants

Plant	Substances	Uses
Gopher plant (Euphorbia)	Latex Sterols	Rubber, petroleum substitutes, drugs
Jojoba (Simmondsia)	Long-chain esters	Surfactants, emulsifiers, waxes, lubricants, preservatives, cosmetics
Buffalo gourd (Cucurbita)	Starch Linoleic acid	Sweeteners Edible oils
Guayule (Parthenium)	Latex	Rubber
Scorpion flower (Phacelia)	Latex Chromenes	Rubber Insecticides
Milkweed (Asclepias)	Latex Silk-like fiber	Rubber, chemical feedstocks, textiles
Juniper (Pinaceae)	Terpenoids Cadinene	Antimicrobials Insecticides
Varthemia candicans	Sesquiterpene lactones	Antimicrobials
Jotropha	Vegetable oils	Surfactants
Meadowfoam (Limnanthes)	Fatty acids	Surfactants
Money plant (Lunaria)	Fatty acids	Surfactants Lubricants
Bladderpod (Lesquerella)	Hydroxy fatty acids	Lubricants Ointments
Thistle (Chamaepeuce)	Hydroxy fatty acids	Lubricants
Kinkaoil ironweed (Vernonia)	Epoxy fatty acids	Plastics Coatings
Hartleaf Christmasbush (Alchornea)	Epoxy fatty acids	Coatings

Fermentation technology is not new to the chemical industry. Prior to World War II (before the introduction of cheap oil), scores of chemicals were manufactured by fermentation processes. For example, only 36 percent of total ethanol production during the mid-1940s was based on petroleum sources; the remainder was made biologically. However, ten years later, almost 60 percent of the ethanol production was derived from oil. Fumaric acid was also manufactured on a commercial scale by fermentation. Production by this route ceased when a more economical synthesis from benzene was developed. In general, once a chemical process using petroleum was developed, it quickly replaced the existing fermentation process.

In spite of this history, a few chemicals are now produced by fermentation, notably citric acid, lactic acid, and various amino acids. These processes have all been improved over the years using applied genetics (e.g., microbial mutagenesis), but recombinant DNA technology has yet to have an impact in this area.

Citric acid is the most important acidulant in the food industry, representing 55-65 percent of the acidulant market. This acid also has pharmaceutical and chemical processing applications. Citric acid is produced commercially using the fungal organism, *Aspergillus niger*. The efficiency of this mold has been dramatically improved using mutagenic techniques. A four-fold increase in product yield has been obtained.

The bacterium *Lactobacillus* is used in the commercial production of lactic acid. Large quantities of this product are obtained using such raw materials as sucrose, glucose, and lactose (from cheese whey). Most of the problems in the manufacture of lactic acid exist in product recovery, not in the fermentative process itself. Thus far, biotechnology has been applied very little to improve this industrial process.

World production of amino acids is currently dominated by Japan; there is very little domestic U.S. production. The bulk of amino acids production is destined for research applications and to nutritional or biomedical preparations. Three amino acids are particularly useful: glutamic acid for the production of monosodium glutamate (MSG), a flavor enhancer; lysine and methionine as animal feed additives.

Glutamic acid production provided the first instance in which biotechnological methods were applied to enhancing amino acid production. The method involves the manipulation of microbial growth conditions and isolating mutant strains. Glutamate is produced in the presence of ammonia by a species of *Corynebacterium*. Growth of this particular species also requires the addition of biotin to the growth medium. In the presence of low concentrations of biotin, bacterial cell membranes become leaky to small molecules, thereby permitting glutamate to diffuse out of the cell. But at high biotin levels, the membranes are normal and prevent glutamate secretion. Furthermore, the biosynthesis of glutamate is reduced in the presence of high biotin levels through a feedback inhibition mechanism.

Lysine is produced both by chemical and fermentation processes. This represents one example where the chemical production method has not totally replaced the biological procedure. Due primarily to the lower direct operating costs incurred by fermentation procedures, about 80 percent of the lysine production world-wide in 1980 was via microbial means. The United States imported about 7,000 metric tons of lysine in 1979. The Japanese firm, Ajinomoto, recently patented a recombinant strain of *E. coli* that contains multiple copies of lysine-producing genes encoded on plasmids.

Recent announcements made by Bethesda Research Laboratories (BRL) indicate that recombinant DNA technology has been used to isolate some of the genes required in the synthesis of the amino acid proline. BRL is currently seeking ways to exploit this discovery on an industrial scale.

Table 4.4 lists those amino acids that are produced microbiologically and the bacterial species used in their manufacture.

Table 4.4: Fermentative Production of Amino Acids from Glucose

Amino Acid	Yield (g/ℓ)	Microorganism
DL-Alanine	40	*Corynebacterium gelatinosum*
L-Arginine	29	*Brevibacterium flavum*
L-Citrulline	30	*Brevibacterium flavum*
L-Histidine	10	*Brevibacterium flavum*
L-Homoserine	15	*Corynebacterium glutamicum*
L-Isoleucine	15	*Brevibacterium flavum*
L-Leucine	28	*Brevibacterium lactofermentum*
L-Lysine	32	*Brevibacterium flavum*
	44	*Corynebacterium glutamicum*
L-Ornithine	26	*Corynebacterium glutamicum*
L-Phenylalanine	2	*Brevibacterium flavum*
	6	*Bacillus subtilis*
L-Proline	29	*Brevibacterium flavum*
L-Threonine	18	*Brevibacterium flavum*
L-Tryptophan	2	*Brevibacterium flavum*
L-Valine	23	*Brevibacterium lactofermentum*

Future Prospects

A variety of issues relevant to the future of biotechnology in the chemical industry can be adduced:

> It is unlikely that biological processes will be applied in the near future to the large-scale manufacture of most commodity chemicals; i.e., bulk chemicals whose production capacity is measured in the millions of pounds annually. Although many of these products are derived from ever-more-costly petroleum feedstocks, bioprocesses will be unable to compete economically with traditional synthetic routes for 10-20 years or more. There exists obvious exceptions to this general conclusion, such as ethanol and some short-chain organic acids (Table 4.2), but even these substances will be more cheaply produced by conventional methods for some time to come.
>
> A significant role for applied genetics in the chemical

industry will be in the manufacture of high-priced specialty chemicals or in synthesizing new chemicals that have no practical alternative route. Enzymes will be employed as highly specific catalysts for performing discrete chemical steps in a synthetic route. Microorganisms that express the desired enzyme activity may be used directly. Microbes will be sought that carry out chemical transformations otherwise requiring large inputs of energy, such as hydrogenations, amidation, etc.

The economics favoring the use of bioprocesses in the chemical industry will depend substantially on process design and engineering characteristics, rather than on the biotechnology involved. This is true for the chemical industry to a much greater extent than for the pharmaceutical industry. Thus, practical applications of biotechnology in this industrial sector will appear slowly and only following extensive analysis of the relevant biochemical engineering factors.

ENERGY INDUSTRY

Current Activities

The potential applications of biotechnology in the energy field are vast. Proponents anticipate that a sizeable proportion of future world energy needs will be met through biological processes. Genetic engineering of microbes and higher plants will undoubtedly have significant impact on the development of future bioenergy systems, although activities to date have shown little evidence of this practice. Current activities in this industry will be considered within two general areas: energy from biomass and enhanced oil recovery.

Energy from Biomass: Biomass resources encompass all the storage repositories of solar energy. This included photosynthetic organisms of all types, organisms that feed on photosynthetic biomass, and animal wastes. Biomass is a renewable energy source, a quality that distinguishes it from fossil fuels, which are also derived from biomass, but which require eons of time to develop. The energy content of the carbohydrates generated annually in higher plants alone has been estimated to be ten times the global energy consumption. The inclusion of marine biomass, such as phytoplankton, might increase this factor another ten-fold. Clearly, tapping this vast energy supply must be considered a top priority in the years ahead.

Biomass fuel sources have several major advantages over fossil fuels:

They are renewable, flexible through crop switching, and adaptable through genetic manipulation.

They do not contribute to carbon dioxide pollution because, at a steady state, carbon dioxide is incorporated into plant material and removed from the atmosphere at the same rate that it is put into the atmosphere by combustion.

The rate of carbon dioxide fixation into usable plant material by photosynthesis is fifty times greater than our current rate of fossil fuel consumption.

Biomass potential is more evenly distributed geographically than are fossil fuel reserves.

The potential market in biomass is huge allowing R&D costs to be amortized over a large number of production units.

Biological systems useful in the conversion of biomass to liquid fuel have not been intensely developed. Current commercial practice is founded on the production of alcohol for distilled beverages. Corn is the main feedstock and the yeast *Saccharomyces cerevisiae* is the principal fermentation organism. It is clear that *S. cerevisiae* can be made to convert carbohydrates by fermentation to ethanol with a much higher efficiency than is currently achieved. This higher yield can be approached in two ways: (1) a greater mass of ethanol can be produced per mass of carbohydrate consumed, and (2) a product with a higher percentage of ethanol can be produced. The overall efficiency of the process can be improved by exploring mixed bacterial-yeast fermentation systems and by adapting the whole fermentation process to a continuous flow mode.

Ethanol for use as fuel, either alone or mixed with gasoline to make gasohol, can be produced by microbial fermentation of sugars. Two sources of sugars abound. First, starch (for which fermentation technology is well advanced) is available from edible plant products, such as corn, wheat, potatoes, sugar cane, sugar beets, and cassava. Second, cellulose, from which conversion to ethanol is difficult, is abundant in municipal/agricultural wastes and forests. Utilizing starch as the feedstock for ethanol production entails a diversion of crop land that could otherwise contribute to the food supply. This disadvantage has not deterred the government of Brazil from investing $5 billion during the past decade on facilities to manufacture ethanol from cassava, sugar, and molasses. Ultimately, all of Brazil's motor vehicles will be run on ethanol. It is estimated that 2% of the nation's land will be devoted to this enterprise.

Ethanol represents one of the most promising alternative fuels to

OPEC oil. It can be burned in conventional automobile engines without modification as a 20% alcohol/80% gasoline mixture. Relatively minor engine and fuel system adjustments are required to convert gasoline engines to 100% ethanol use. The Ford Motor Co. of Brazil currently sells a conversion kit for about $250 that will convert a standard auto engine to permit use of 100% ethanol. Alcohol can be efficiently handled as a fuel by existing petroleum distribution networks.

Ethanol production by yeast may be greatly enhanced using molecular cloning techniques. The biochemical pathway unique to ethanol metabolism is relatively simple. Pyruvate is converted by the enzyme pyruvate decarboxylase to acetaldehyde and carbon dixoide. Acetaldehyde is converted to ethanol by the enzyme alcohol dehydrogenase. The gene for alcohol dehydrogenase has been cloned in several laboratories and it appears possible to increase the efficiency of the fermentation process by increasing the level of alcohol dehydrogenase in the cell using genetic engineering techniques. The other enzyme in the process, pyruvate decarboxylase, should also be amenable to genetic engineering.

In the long run, ethanol production from cellulosic wastes will be preferable to using foodstuffs as the raw material. Typical cellulosic materials consist of 50% cellulose (a glucose polymer), 25% hemicellulose (a polymer of xylose, a five-carbon sugar), and 25% lignin (a complex phenolic polymer). This semicrystalline lignocellulose is broken down with difficulty into fermentable constituents, by acid treatment or by the enzyme cellulase. This expensive initial phase of cellulose preparation is where process improvements are most needed. In a recent advance reported by a new firm, the Biological Energy Corp. (80%-owned by General Electric), thermophilic bacteria from the *Thermomonospora* species are able, in a process operated at 55°C, to convert cellulose to glucose with 93% efficiency. Figure 4.4 provides a general scheme for ethanol production utilizing either cellulosic or starch feedstocks, and Figure 4.5 provides an overview of the variety of petrochemical feedstocks that can be obtained from cellulosic starting materials.

One drawback in utilizing *Saccharomyces* strains for ethanol production relates to its strict preference for 6-carbon sugars (such as glucose) as fermentation substrates. Scientists at the USDA's Northern Regional Research Laboratory in Illinois recently uncovered a yeast, *Pachysolen tannophilus*, that readily converts both glucose and xylose to ethanol. Since xylose represents a major fraction of cellulosic biomass, further examination and improvement of this newly discovered strain promises to greatly enhance ethanol yields from unconventional fermentations.

Methane generation by anaerobic digestion of biomass provides another route whereby renewable resources are utilized for energy production. Animal feedlot wastes and municipal sewage are most often cited as providing the raw materials for this process, although forest residues and food crop biomass are also suitable. The process produces:

Biogas, consisting of approximately 60% methane and 40% carbon dioxide.

Residual solids, containing vegetable proteins, which have potential value as feed additives or fertilizers.

Spent process water, laden with nutrients, which is suitable for growing algae or as a fertilizer.

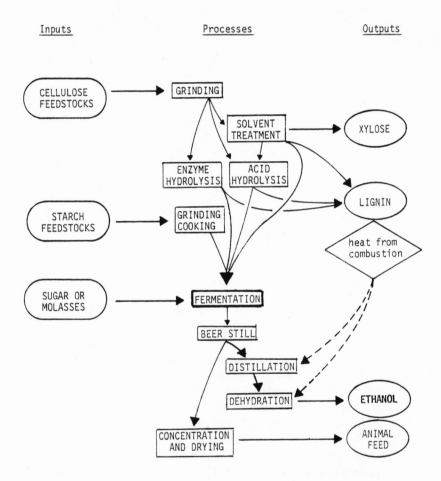

Figure 4.4: Steps in the conversion of biomass to ethanol and by-products. (Source: King, S.R., 1979.)

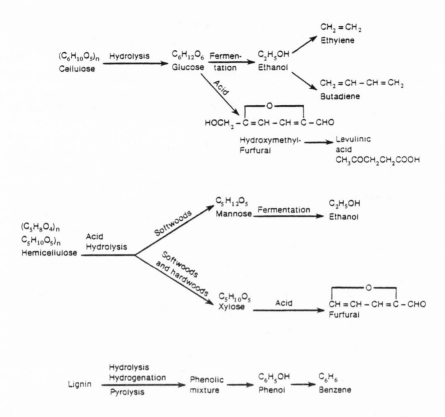

Figure 4.5: The conversion of lignocellulose into useful chemical feedstocks.

Since the process occurs in closed digesters to exclude oxygen, the waste materials used as feedstock are prevented from spoiling the environment or giving rise to pathogenic organisms. Anaerobic reactors are classified into three types depending on the operating temperature (i.e., the optimal temperature for growth of the particular microbial strain involved): (1) psychophilic (under 20°C), (2) mesophilic (20°C to 45°C), and (3) thermophilic (45° to 65°C). A typical digester used for sewage treatment is depicted in Figure 4.6.

Most applications of this technology involve small, community-scale operations. Biogas generators associated with large animal feedlots or municipal sewage treatment facilities might readily supply the energy needs of the local population. Simple anaerobic digesters of this type are common in the People's Republic of China, Korea, Taiwan, and India. But large-scale operations may be feasible. A study commissioned by the U.S. Science and Education Administration of the USDA found that economical biogas production could be achieved with feedlots averaging

1,000-2,000 head of cattle in size. Others have proposed a large, centralized facility that could produce 50 million cubic feet of methane per day using biomass crops as feedstocks. Also, the mass-cultivation of water hyacinths on sewage lagoons for use as fermentation substrate has been proposed.

Current R&D efforts in this area tend to emphasize aspects of process design and engineering rather than microbiology. Nevertheless, genetic engineering may have a significant role in future developments of biogas production. Microbial methanogenesis typically occurs in two stages, with different bacterial populations responsible for each. In the first, or acidogenic, stage the cellulosic substrate is broken down into various organic acids, causing a sharp drop in reactor pH. In the second, or methanogenic, stage the acids are converted to methane and carbon dioxide. As frequently occurs, overactive acidogenic microbes lead to a drop in pH sufficient to inhibit the activity of the methanogenic organisms–a condition known as digester "souring" in which methane production ceases. It may be possible through genetic engineering to combine the acidogenic and methanogenic activities in a single microbial population so that the proper balance of activities is maintained irrespective of population size.

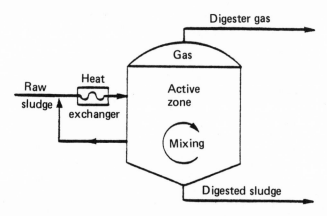

Figure 4.6: A single-tank anaerobic digester of biogas production. (Source: U.S. Environmental Protection Agency, 1979.)

Hydrocarbons are synthesized and accumulated by a wide variety of bacteria, algae, and yeasts (see Table 4.5). These microbes generally extract carbon dioxide from air and utilize energy derived from photosynthesis to reduce chemically and polymerize CO_2 into long-chain lipids. Some microbes utilize carbohydrates such as glucose as carbon sources. As much as 40-50% of the dry weight of certain oil-bearing microorganisms can consist of reduced hydrocarbon materials suitable as substitute fuels. Many higher plant species produce a sap of fruiting body that is high in hydrocarbon content. Most familiar are vegetable oils, such as sunflower, cottonseed, linseed, palm, etc., some of which are under investigation as diesel fuel additives. A variety of less familiar tropical plants and trees is also under examination as hydrocarbon producers.

Table 4.5: Some Species of Algae That Produce Hydrocarbons

Species	Lipid Content (% dry wt)
Biddulphia aurita	12.2
Chlamydomonas applanate	32.8
Chlorella pyrenoidosa	14.4
Chlorella vulgaris	28.8
Monallanthus salina	40.8
Nannochloris sp.	20.2
Nitzschia palea	22.2
Oocystis polymorpha	12.6
Ourococcus sp.	27.0
Skeletonema costatum	23.8

Source: Shifrin, N.S. and Chisholm, S.W. (1980) in *Algae Biomass*, p. 633, Elsevier Press.

The production of hydrogen gas from water has been demonstrated in laboratory studies—its commercial-scale feasibility remains to be shown. The system utilizes units of photosynthetic activity, called chloroplasts, isolated from green plants, such as lettuce or spinach. A biophotolysis reaction is established in which energy from sunlight splits water (H_2O) into molecular hydrogen (H_2) and oxygen (O_2). Successful operation of the system requires a means of removing oxygen to prevent reaction with hydrogen to regenerate water. Rather than isolating chloroplasts from higher plants, it may be preferable to use intact, photosynthetic algae or bacteria. The future use of hydrogen as a fuel offers the promise of a nonpolluting, inexhaustible energy source. However, numerous technical obstacles remain before this prospect will be realized.

Enhanced Oil Recovery: A second major area of the energy field in which applied genetics will have an impact is not the creation of new sources of energy but enhanced recovery from existing energy supplies. Primary and secondary oil recovery techniques manage to extract only about one-half of a known oil reservoir. An estimated 200 billion barrels of oil in the continental United States remain out of reach with conventional recovery techniques. A variety of microbial-based tertiary recovery methods has been proposed as a means to tap this vast resource. These include:

The injection of oil-degrading bacteria into an oil field would reduce the oil's viscosity, or convert oil to natural gas.

The injection of microbes to repressurize a spent oil well by synthesizing carbon dioxide or other gaseous metabolite.

The injection of microbes that manufacture and secrete chemical surfactants that would act to mobilize tightly bound oil.

In addition to these potential applications in existing oil fields, microbial processes have been promoted for use in extracting tar and oil (bitumen) lodged in tar sands. Also, a bacterial process is under development at the University of Southern California that would release kerogen (a petroleum material) from oil shale. This process could generate a barrel of oil per ton of western oil shale without extensive ore crushing, retorting, or environmental damage that attend strictly physical recovery methods. Finally, analysis of subsoil microbial populations may assist in locating previously unknown oil and gas fields below. This microbiological method of prospecting for petroleum is still in an early stage of commercial development, as are all of the microbial recovery methods described above. Moreover, the physical and chemical characteristics of oil reservoirs differ greatly one from another, and conditions in any oil well–such as high temperatures, high salt and sulfur concentrations, extremes of pH–are rarely conducive to optimal growth of microorganisms.

Future Prospects

The range of potential uses of applied genetics in the energy industry appears to be far wider than in the chemicals sector. However, most of these possibilities lie far in the future, at least with regard to large-scale commercial application. Development of systems for ethanol production from biomass for use in gasohol are proceeding apace, especially in petroleum-poor areas like Brazil, but the economics of this process and the energy saving incurred will remain unfavorable, probably for the remainder of the 1980s. Nevertheless, several long-range projects can be envisioned that may one day provide significant sources of energy.

Mass production of hydrocarbon substances from various species of higher plants, including those listed in Table 4.3, can become economically feasible when either (1) plant cells are manipulated to grow in massive cultivators, akin to microbial fermentors, in which excreted hydrocarbons are continuously collected, or (2) the genetic information that enables the plant cell to synthesize hydrocarbons is transferred to microorganisms which, in turn, manufacture and excrete the fuel-like substances. The biotechnical and engineering obstacles that stand in the way of such a project are formidable.

A biological solar battery will someday replace the panels of silicon solar cells that find specialized uses today. The biological battery will operate via a direct conversion of sunlight into electricity (i.e., a current of electrons) that is generated during photosynthesis. Although all green plants engage in photosynthesis and are, therefore, suitable sources of materials for constructing a biological battery, a primitive, purple photosynthetic bacterium, called *Rhodospirillum rubrum,* may be exploited as the living solar cell. Alternatively, the photosynthetic blue-green algae, which utilize carbon dioxide and nitrogen directly from air may serve this purpose.

Ethanol production may become more efficient through use of microorganisms other than common yeasts (e.g., *Saccharomyces cerevisiae,* or brewer's yeast). A bacterial species, called *Zymomonas mobilis,* carries out alcoholic fermentation two to three times faster than yeasts. This bacterium, now employed to make tequila, is under investigation by researchers at the USDA's Northern Regional Research Lab in Peoria, Illinois.

Acidophilic, iron-oxidizing *Thiobacilli* bacteria may prove useful in oil shale or coal conversion processes. The bacteria will mobilize the inorganic mineral content of the shale or coal without affecting the hydrocarbon content of the material. The porous zones that this process generates *in situ* may assist in subsequent retorting or gasification schemes.

MINING INDUSTRY

Current Activities

The impact of biotechnology on the mining industry is currently quite limited in scope, consisting of two general areas:

The accumulation of metals by organisms, either by binding at cell surfaces or by intracellular uptake of metals.

Biochemical transformations of metals, including solubilization or precipitation, oxidation/reduction processes, and the interconversion of inorganic and organic metal compounds.

The various bioprocesses subtended under these categories are all carried out by a relatively small number of bacterial species. Figure 4.7 lists these organisms and summarizes the means by which these microbes extract energy from chemically reduced inorganic compounds (such as ferrous iron or sulfur compounds) and employ either inorganic (CO_2) or organic carbon sources.

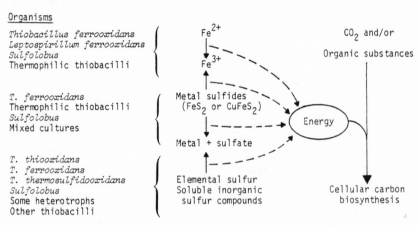

Figure 4.7: Leaching bacteria: organisms and basic metabolism. (Source: Bull, A.T., et al., 1979.)

The microbial process whereby metals are solubilized from their ores is called bacterial leaching. The operation consists of percolating acidified water through heaps or dumps of low-grade ore that may contain up to four billion tons of rock. Bacterial action within the dump oxidizes mineral sulfide, producing sulfuric acid, and solubilizes the metal. The solution, or leachate, is collected and processed to recover the dissolved metal. The residual liquid, containing sulfuric acid and ferrous/ferric iron, is recycled to the dump. This somewhat crude process has been used in mining operations since Roman times. Currently, its greatest use occurs in copper and uranium mining operations. Approximately 12% of U.S. copper production stems from dump leaching of this type.

The feasibility of large-scale dump leaching was first demonstrated in 1750 in Rio Tinto, Spain. The technology is still practiced widely in mining operations throughout the western United States, including mines owned by Kennecott Copper at Bingham Canyon, Utah and Santa Rita, New Mexico, as well as the Butte, Montana mines operated by Anaconda Copper.

Despite the long history of mineral leaching, the role of microorganisms as mediators of the process was not recognized until the mid-1950s. The principal microbes involved in copper extraction are *Thiobacillus ferrooxidans* and *Thiobacillus thiooxidans*. Both species are rod-shaped, aerobic bacteria that thrive in an acid environment (pH 1.5 to 3.0) and use carbon dioxide as a carbon source. They function within a temperature range of 18° to 40°C (64° to 104°F).

The bacteria require, in addition to water and oxygen, a reduced iron or sulfur energy source, as seen in the following equations (unbalanced):

$$Fe^{2+}SO_4 \ + \ O_2 \ + \ H_2SO_4 \ \longrightarrow \ Fe_2^{3+}(SO_4)_3 \ + \ H_2O$$

$$S_8 \ + \ O_2 \ + \ H_2O \ \longrightarrow \ H_2SO_4$$

$$H_2S \ + \ O_2 \ \longrightarrow \ H_2SO_4$$

Ferric iron (Fe^{3+}) and sulfuric acid (H_2SO_4) generated by these bacterial reactions are very effective chemical solubilizers for numerous minerals, including those listed in Table 4.6. In addition to those shown, other minerals, such as uranium oxides, that co-exist with iron or sulfur-containing ores, are readily leached. These reactions occur at rates approximately 500,000-fold faster than the oxidation of iron and sulfur by air in the absence of bacteria. Both of the *Thiobacillus* species are found in great abundance in leaching operations–as many as 10^7 organisms per gram of ore. Indeed, the high concentration of microbes in leachate solutions poses difficulties during subsequent mineral extraction and isolation.

Table 4.6: Minerals Readily Leached by Bacterial Action

Mineral	Formula
Pyrite	FeS_2
Chalcopyrite	$CuFeS_2$
Chalcocite	Cu_2S
Covellite	CuS
Arsenopyrite	$AsFeS$
Molybdenite	MoS_2
Stibnite	Sb_2S_3
Pentlandite	$NiFeS_2$
Zincblende	ZnS

Bacterial leaching is also utilized to recover uranium from low-grade ores, mine tailings, and other ores that are rich in pyrite. The following reaction pertains to this process:

$$UO_2 + Fe_2^{3+}(SO_4)_3 \longrightarrow UO_2SO_4 + Fe^{2+}SO_4$$

This solubilization process has been used in scavenger operations in mined-out and low-grade stopes in the Elliot Lake region of Ontario. The high efficiency of bacterial leaching of uranium has led to the frequent use of *in situ* mining in Canadian ore beds. Also called "underground solution mining," this process entails the percolation of water through mine shafts and subsequent pumping of the leachate to the surface. Though slower, this approach confers significant safety and environmental benefits. The ores of northern Ontario are amenable to bacterial leaching due to the presence of large amounts of pyrite, whereas the uranium deposits in the U.S. Rocky Mountains and southern Texas contain insufficient pyrite to allow successful leaching operations.

Other bacterial species have been implicated in mineral leaching, including some members of the *Sulfolobus* genus. These bacteria are obligate thermophiles, requiring a temperature range of 45° to 80°C (110° to 175°F), as well as an acidic environment. *Leptospirillum ferrooxidans* is another iron-oxiding acidophile that has been shown to release pyrite more efficiency than *T. ferrooxidans* when grown in mixed cultures with sulfur-oxidizing bacteria.

All organisms, including microbes, can accumulate certain metal ions that are essential for metabolic acitivity. Iron, magnesium, zinc, manganese, copper, cobalt, nickel, moybdenum, and vanadium are required by various organisms, albeit frequently in trace quantities only. Nevertheless, certain microbes have evolved highly efficient means of permitting the selective concentration of metals far in excess of the local concentration. Toxic metals, such as cadmium, lead, silver, and thallium, can be accumulated even though these substances have no metabolic function. Apart from the intracellular uptake of metals, positively charged metal ions can be removed from solution by adsorption onto the negatively charged surface of the microbe.

Microorganisms can also be utilized in the restoration of wastewaters from mining and milling operations. One successful operation uses algae to remove both soluble and particulate lead from the mill tailings of several mining ventures in Missouri.

These operations consist of settling ponds and a series of shallow meanders in which the algae are encouraged to grow. Chemical analysis has shown that the algae accumulate heavy metals from the effluent released from the settling pond. Algae species that have been identified to function effectively in these types of operations include: *Cladophora, Rhizoclonium, Hydrodictyon, Spirogyra, Potamogeton,* and *Oscillatoria.*

At present the potential to use genetic engineering to improve the performance of these, or other, algae species is remote, and may not be possible for several years. On the other hand, genetic engineering techniques could certainly be used to improve *Thiobacillus ferrooxidans*.

As explained above, *T. ferrooxidans* derives its energy from the oxidation of ferrous ion, metal sulfides, and soluble sulfur compounds in an acidic medium. The ferric ion generated in the form of ferric sulfate is then able to react chemically with several ore minerals and oxidize them. The ferric ion is then regenerated by the microorganisms. Apparently, one of the primary rate-limiting steps in the leaching of metal ores is the ferrous-to-ferric reoxidation. Ferric ion competitively inhibits the rate of ferrous ion oxidation. Thus, as the concentration of ferric ion increases, its production is slowed.

Since ferric ion has no other metabolic effect on *T. ferrooxidans* except to slow its own production, it should be straightforward to isolate a suitable mutant strain that is not affected by ferric ion concentration.

Thiobacilli are able to develop considerable resistance to the very high concentrations of the metals being leached, but the microbe is inhibited by some metals such as silver, mercury, and cadmium, at quite low concentrations. Metabolic resistance to heavy metals is frequently conferred by the presence of certain bacterial plasmids. Experiments could be undertaken to isolate appropriate plasmids from other bacteria and to introduce them into *Thiobacilli* using recombinant DNA or conventional genetic technologies.

Improved bacterial growth and mineral leaching activity have resulted when *Thiobacilli* are grown in conjunction with the nitrogen-fixing bacterium, *Beijerinckia lacticogenes*. This latter bacterial species probably supplies *Thiobacilli* with essential nitrogenous nutrients. Since *B. lacticogenes* is less able to withstand the highly acidic environment required by *Thiobacilli*, it may prove worthwhile to introduce the nitrogen fixation genes (*nif* genes) directly into *Thiobacilli*. Alternatively, *nif* genes from *Azotobacter* or *Klebsiella* (two free-living nitrogen fixers) can be utilized since these species share structural and biochemical features with *Thiobacilli*.

Future Prospects

Of the industrial sectors considered in this report, the mining industry has demonstrated the least interest in applying biotechnology to its operations. The types of bioprocesses that do pertain to mining are rather limited in scope, but technical advances, leading to increased interest on the part of the industry, can be envisioned.

> All known strains of leaching bacteria are aerobic; that is, they require oxygen. However, essentially oxygen-free conditions exist in the center of huge slag heaps of

low-grade ore. Thus, the engineering of anaerobic strains of *Thiobacillus* would be received with great enthusiasm by the mining industry. The technical feasibility of this proposal is uncertain. Likewise, development of improved thermophilic leaching bacteria would be very useful, owing to the heat generated within ore dumps.

The United States relies on imports for the vast majority of certain mineral resources, including chromium, titanium, and manganese (see Table 4.7). Recycling of these materials is of increasing importance. The development of efficient microbiological systems for extracting these metals from industrial effluents and other waste repositories would constitute a major industrial and political tour de force.

Very little basic information is available regarding the biochemistry and genetics of leaching bacteria. Consequently, genetic engineering, especially recombinant DNA, will have little impact on developments in this area for at least five years. The properties and commercial suitabilities of existing, naturally occurring leaching bacteria will undergo thorough examination first.

Table 4.7: Strategic Minerals and U.S. Dependence on Foreign Sources

Mineral	Uses	Percentage Imported	Sources
Bauxite	Aluminum	94	Jamaica, Guinea, Surinam
Chromium	Ferroalloys	91	South Africa, USSR
Cobalt	Superalloys	93	Zaire, Belgium, Zambia
Columbium	Ferroalloys	100	Brazil
Manganese	Steel	97	Gabon, South Africa
Nickel	Steel	73	Canada
Platinum	Catalysts	87	South Africa, USSR
Rutile	Pigments	100	Australia
Tantalum	Electronics components	97	Thailand
Titanium	Aerospace components	47	Japan, USSR

POLLUTION CONTROL INDUSTRY

Current Activities

The organic matter invested in all living things whether plant or animal, is eventually recycled back into the environment as CO_2. The process whereby organic carbon is converted into inorganic carbon is called mineralization. Representing a major portion of the overall carbon cycle (see Figure 4.8), mineralization is almost always a consequence of microbial action. That is, bacterial decomposers are ultimately responsible for the degradation of all organic carbon-containing substances in the biosphere. For example, bacteria of the *Pseudomonas* species metabolize simple alkane compounds, such as octane, by means of an enzymatic oxidation pathway that converts the alkane (R-CH$_3$) into the corresponding carboxylic acid (R-COOH). The acid is then consumed as an energy source by the bacterium via further oxidation to carbon dioxide.

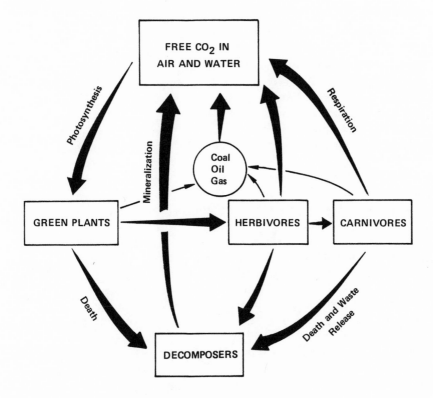

Figure 4.8: The carbon cycle.

Therefore, it is hardly surprising that microbiology plays an important role in pollution control and waste management, particularly in the case of organic pollutants. Moreover, inorganic pollutants, such as nitrogen-containing substances and toxic metals, are often treatable using biological systems. Biological waste management has been practiced by mankind literally for thousands of years, but modern advances in applied genetics may revolutionize the pollution control industry to the extent that bioprocesses may soon replace many currently employed chemical/physical systems. Current activities within this industry fall under three general headings:

> Biodegradation of organic substances, such as petroleum products, pesticides, herbicides, industrial solvents, and lignin wastes.
>
> Biological denitrification and desulfurization processes.
>
> Removal or concentration of toxic heavy metals.

Each of these areas will be examined in the following sections.

Biodegradation of Organic Substances: Most current activity, and that which has the most potential for biological waste treatment, exploits the capacity of microbes to degrade toxic chemical pollutants. A great variety of naturally occurring microorganisms, largely isolated from soil or aquatic environments, are known to utilize hazardous organic substances as carbon and energy sources. Table 4.8 provides a sample of the biodegradative processes that are currently in use or under investigation. Figure 4.9 shows the degradative pathways of several specific pollutants.

Efforts to improve on nature by applications of genetic engineering in this area have been minimal to date, but such activities are certain to increase in the near future. Several efforts bear mentioning. Among the first highly publicized uses of genetic engineering was that of Chakrabarty, then at General Electric. He combined the qualities of several strains of the bacterium *Pseudomonas,* each of which could degrade a single hydrocarbon component of crude oil, into a single bacterial strain. This "man-made" bacterial culture proved superior to a mixed microbial culture composed of each of the contributing strains in breaking down crude petroleum. The "oil-eating" microbes feed on the crude petroleum, converting the hydrocarbon compounds into cellular constituents (biomass) and carbon dioxide. However, the petroleum constituents that are converted by this microbial strain (namely camphor, naphthalene, and short-chain alkanes) are not the major environmental concerns arising from oil spills. These volatile hydrocarbons are either vaporized or readily degraded by natural bacterial action. Of greater impact are the various asphaltenes that constitute the heavy, nonvolatile fraction of crude oil. These compounds are extremely refractory to microbial degradation. Despite the publicity that attended Chakrabarty's efforts, GE has not pursued this project beyond the laboratory stage of development.

Table 4.8: Microbial Degradation of Various Organic Pollutants

Pollutant	Microbes Involved
I. Petroleum hydrocarbons	200+ species of bacteria, yeasts, and fungi; e.g., *Acinetobacter*, *Arthrobacter*, *Mycobacteria*, *Actinomycetes*, and *Pseudomonas* among bacteria; *Cladosporium* and *Scolecobasidium* among yeasts
II. Pesticides/herbicides	
Cyclodiene type (e.g., aldrin, dieldrin)	*Zylerion xylestrix* (fungus)
Organophosphorus type (e.g., parathion, malathion)	*Pseudomonas*
2,4-D	*Pseudomonas, Arthrobacter*
DDT	*Penicillium* (fungus)
Kepone	*Pseudomonas*
Piperonylic acid	*Pseudomonas*
III. Other chemicals	
Bis(2-ethylhexyl)phthalate	*Serratia marascens* (bacteria)
Dimethylnitrosamine	Photosynthetic bacteria
Ethylbenzene	*Nocardia tartaricans* (bacteria)
Pentachlorophenol	*Pseudomonas*
IV. Lignocellulosic wastes	
Municipal sewage	*Pseudomonas* *Thermonospora* (a thermophilic bacterium)
Pulp mill lignins (various phenols)	Yeasts: *Aspergillus* *Trichosporon* Bacteria: *Arthrobacter* *Chromobacter* *Pseudomonas* *Xanthomonas*

More recently, Chakrabarty and his colleagues at the University of Illinois have engineered the development of bacteria that degrade 2,4,5-T, a ubiquitous, persistent, and highly toxic herbicide commonly known as Agent Orange. In work funded by the U.S. Environmental Protection Agency, these scientists co-cultured microbial strains isolated from herbicide-contaminated soils with strains of *Pseudomonas* known to contain plasmids encoding various biodegradative activities. The mixed microbial culture was adapted (over a period of 8 to 10 months) to grow in the presence of increasing concentrations of 2,4,5,-T. Chakrabarty predicts that heavily contaminated areas, such as those laid to waste over 15 years ago by U.S. Air Force target practice with Agent Orange, can now be cleaned up in a matter of weeks. Only such field trials will verify this claim, as well as assess any residual effects of releasing large amounts of genetically altered bacteria into the environment.

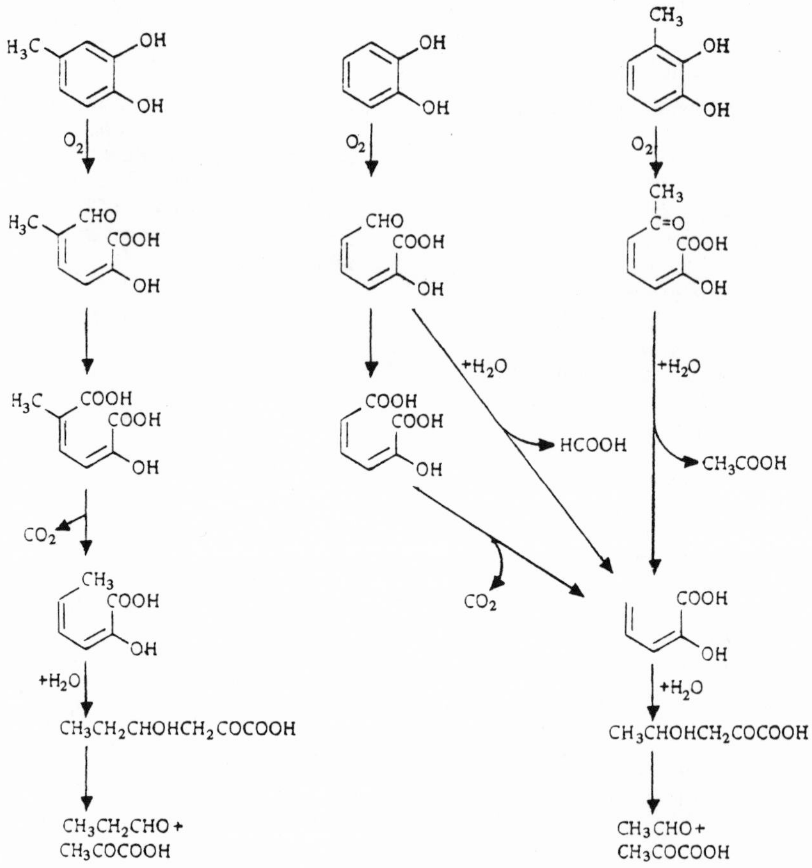

Figure 4.9: Degradation pathways of several phenolic compounds by *Pseudomonas putida*. (Source: Bull, A.T., et al., 1979.)

Scientists at the Battelle Memorial Institute in Columbus, Ohio, are engaged in genetic engineering of microbes that efficiently degrade the chlorinated herbicides, 2,4-D and atrazine. Likewise, SRI International has undertaken a program to compile a list of common toxic chemicals that are amenable to microbial biodegradation and to isolate and engineer improved strains that might have commercial value.

In general, chlorinated organics are more recalcitrant to biodegradation than are nonchlorinated substances. Thus, persistent pollutants such as DDT, PCBs, etc., represent a more serious challenge to pollution control engineers who are hopeful of applying biological treatment systems to waste management. Microbes exist that can perform chemical transformations of these recalcitrant substances (see Table 4.9), but

Table 4.9: Type Reactions for Transformation of Chemicals of Environmental Importance

Reaction Type	Reaction*	Example
Dehalogenation	$RCH_2Cl \rightarrow RCH_2OH$	Propachlor
	$ArCl \rightarrow ArOH$	Nitrofen
	$ArF \rightarrow ArOH$	Flamprop-methyl
	$ArCl \rightarrow ArH$	Pentachlorophenol
	$Ar_2CHCH_2Cl \rightarrow Ar_2C{=}CH_2$	DDT
	$Ar_2CHCHCl_2 \rightarrow Ar_2C{=}CHCl$	DDT
	$Ar_2CHCCl_3 \rightarrow Ar_2CHCHCl_2$	DDT
	$Ar_2CHCCl_3 \rightarrow Ar_2C{=}CCl_2$	DDT
	$RCCl_3 \rightarrow RCOOH$	N-Serve, DDT
	$HetCl \rightarrow HetOH$	Cyanazine
Deamination	$ArNH_2 \rightarrow ArOH$	Fluchloralin
Decarboxylation	$ArCOOH \rightarrow ArH$	Bifenox
	$Ar_2CHCOOH \rightarrow Ar_2CH_2$	DDT
	$RCH(CH_3)COOH \rightarrow RCH_2CH_3$	Dichlorfop-methyl
	$ArN(R)COOH \rightarrow ArN(R)H$	DDOD
Methyl oxidation	$RCH_3 \rightarrow RCH_2OH$	Bromacil
	$RCH_3 \rightarrow RCHO$	Diisopropylnaphthalene
	$RCH_3 \rightarrow RCOOH$	Pentachlorobenzol
Hydroxylation	$ArH \rightarrow ArOH$	Benthiocarb, Dicamba
	$RCH_2R' \rightarrow RCH(OH)R'$	Carbofuran, DDT
	$R(R')CHR'' \rightarrow R(R')COH(R'')$	Bux insecticide
	$R(R')(R'')CCH_3 \rightarrow$	
	$\quad R(R')(R'')CCH_2OH$	Denmert
β-Oxidation	$ArO(CH_2)_nCH_2CH_2COOH \rightarrow$	ω-(2,4-dichlorophen-
	$\quad ArO(CH_2)_nCOOH + CH_3COOH$	oxy)alkanoic acids
Epoxidation	$RCH{=}CHR' \rightarrow RCH(O)CHR'$	Heptachlor
N-oxidation	$R(R')NR'' \rightarrow R(R')N(O)R''$	Tridemorph
S-oxidation	$RSR' \rightarrow RS(O)R'$ or $RS(O_2)R'$	Aldicarb
=S to =O	$(AlkO)_2P(S)R \rightarrow (AlkO)_2P(O)R$	Parathion
	$RC(S)R' \rightarrow RC(O)R'$	Ethylenethiourea
Sulfoxide reduction	$RS(O)R' \rightarrow RSR'$	Phorate
Triple bond reduction	$RC{\equiv}CH \rightarrow RCH{=}CH_2$	Buturon
Double bond reduction	$Ar_2C{-}CH_2 \rightarrow Ar_2CHCH_3$	DDT
	$Ar_2C{=}CHCl \rightarrow Ar_2CHCH_2Cl$	DDT
Double bond hydration	$Ar_2C{=}CH_2 \rightarrow Ar_2CHCH_2OH$	DDT
Nitro metabolism	$RNO_2 \rightarrow ROH$	Nitrofen
	$RNO_2 \rightarrow RNH_2$	Sumithion
Oxime metabolism	$RCH{=}NOH \rightarrow RC{\equiv}N \rightarrow$	Aldicarb, bromoxynil,
	$\quad RC(O)NH_2$ or $RCOOH$	dichlobenil

*R = organic moiety; Ar = aromatic; Alk = alkyl; and Het = heterocycle.

microorganisms have not yet been isolated that can utilize these compounds as carbon or energy sources. Indeed, it is this lack of direct metabolism by microbes that explains the environmental persistence of compounds such as these. Future success in developing biodegradative systems for pollutants of this type may depend on locating communities consisting of several species of microorganisms which function cooperatively to decompose recalcitrant compounds.

Denitrification and Desulfurization: The various oxidized forms of nitrogen (NO_x) and sulfur (SO_x) present serious environmental concerns owing to the ease with which they are converted to strong acids (e.g., nitric and sulfuric) upon exposure to water. The acidification of lakes and ground water poses a serious threat to the maintenance of aquatic life and fresh water supplies. Although large amounts of nitrogen (and lesser quantities of sulfur) are nutritional requirements for life, the large-scale burning of sulfur and nitrogen-containing fossil fuels and the release of certain industrial wastes have loaded the environment with toxic levels of these inorganic substances. Traditional schemes for reducing the emissions of these pollutants have been largely physical/chemical in design. Biological processes are under investigation, however, and several systems have demonstrated feasibility in laboratory-scale applications.

The biological nitrogen cycle entails three phases involving various oxidation states of nitrogen (see Figure 4.10). Atmospheric nitrogen (N_2) is relatively inert chemically and must be "fixed" into usable forms such as nitrate and nitrite (oxidized nitrogen) or ammonia (reduced nitrogen). Meanwhile, fixed nitrogen is recycled back into the atmosphere by anaerobic processes. All these steps are carried out by various species of bacteria, according to the chemical reactions shown in Figure 4.10. Of these three phases, the third (denitrification) is the least understood, but it is this process that promises to alleviate pollution problems stemming from excess nitrate and ammonia.

Pollution by sulfur-containing compounds presents a more serious problem than pollution by nitrogenous substances because of sulfur's greater toxicity to living organisms and its greater prevalence in fossil fuels and industrial waste streams. Inorganic sulfur compounds, such as sulfate (SO_4^{-2}) and hydrogen sulfide (H_2S), can be metabolized by certain microbial species, as shown in the following reactions:

$$SO_4^{2-} \xrightarrow{\text{I}} H_2S \xrightarrow{\text{II}} S^0$$

I. *Desulfovibrio desulfuricans*
II. *Chlorobium thiosulfatophilum* or *Chromatium vinosum*

Laboratory-scale systems utilizing these microbial populations arranged serially are under investigation for potential use in treating high-sulfur effluents, such as those arising from coal and gypsum mining and general metallurgical operations.

Fossil fuels may contain up to 7% sulfur by weight. This sulfur is generally in one of three forms:

Organic sulfur, in which sulfur is covalently linked to carbon either directly as $R-S-S-R$ or $R-S-R$, or bound as a sulfate, $R-O-SO_3$

Pyritic sulfur in the form of iron pyrite, FeS_2

Inorganic sulfate, SO_4^{-2}

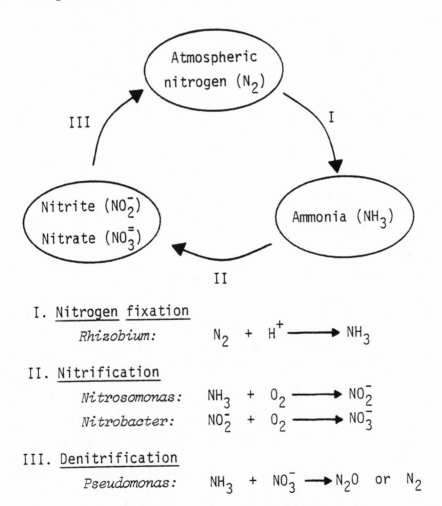

I. Nitrogen fixation

    Rhizobium: $N_2 + H^+ \longrightarrow NH_3$

II. Nitrification

    Nitrosomonas: $NH_3 + O_2 \longrightarrow NO_2^-$

    Nitrobacter: $NO_2^- + O_2 \longrightarrow NO_3^-$

III. Denitrification

    Pseudomonas: $NH_3 + NO_3^- \longrightarrow N_2O$ or N_2

Figure 4.10: Microbial components of the biological nitrogen cycle.

Of these, organic sulfur predominates in crude petroleum, whereas pyritic sulfur and sulfate are found largely in coal. In all cases, the combustion of untreated crude oil or coal releases to the atmosphere huge quantities of sulfur dioxide gas (SO_2) and particulate sulfates. These sulfur compounds are intrinsically toxic and, moreover, combine with water to form sulfuric acid. The removal of sulfur compounds from fossil fuels prior to combustion has been deemed a desirable adjunct to, or possible replacement for, costly scrubbers now widely used to control stack emissions.

Biological desulfurization is still in the experimental stage, but several microbial systems are under investigation. Pyritic sulfur can be leached from mined coal using *Thiobacillus ferrooxidans* or *Thiobacillus thiooxidans*–the same bacterial species employed for mineral leaching in the mining industry. Also, a thermophilic microbe, *Sulfolobus acidocalderius,* has been isolated. All these organisms operate under acidic conditions (pH 1 to 3) and convert sulfides to sulfuric acid. Thus, the pyritic sulfur content of the fossil fuel is transformed into a water soluble compound that can be readily washed away. However, the acid that is generated represents a pollutant in its own right that must be dealt with.

Organic sulfur exists in crude petroleum largely as linear mercaptans (R–SH) or as aromatic thiophenes. Microbial systems for converting thiophenes into water soluble compounds are under development. The biochemistry involved in this transformation is shown in Figure 4.11. The principal drawback of this process lies in the loss of carbon atoms (and, therefore, of Btu content) resulting from the removal of sulfur-containing organics.

Future Prospects

Biological concentration of heavy metal ions from a dilute waste stream involves processes essentially identical to those described for the mining industry. The emphasis for pollution control, of course, is isolation and disposal of toxic metals, whereas ultimate recovery of the metals is of concern to the mining industry. The incorporation of metals from an industrial effluent into biological sediments (i.e., activated sludge) has proven to be a satisfactory application of biotechnology to pollution control. Metals such as aluminum, cadmium, cobalt, nickel, plutonium, and uranium in concentrations from less than 1 ppm to 100 ppm are recoverable. Such dilute solutions are difficult or uneconomical to treat with physical/chemical recovery methods. The immobilization of metals by these sediments may be the result of (1) direct intracellular uptake, (2) adsorption to cell surfaces, or (3) sequestration in a microbially produced exopolysaccharide matrix. In addition to bacteria, other

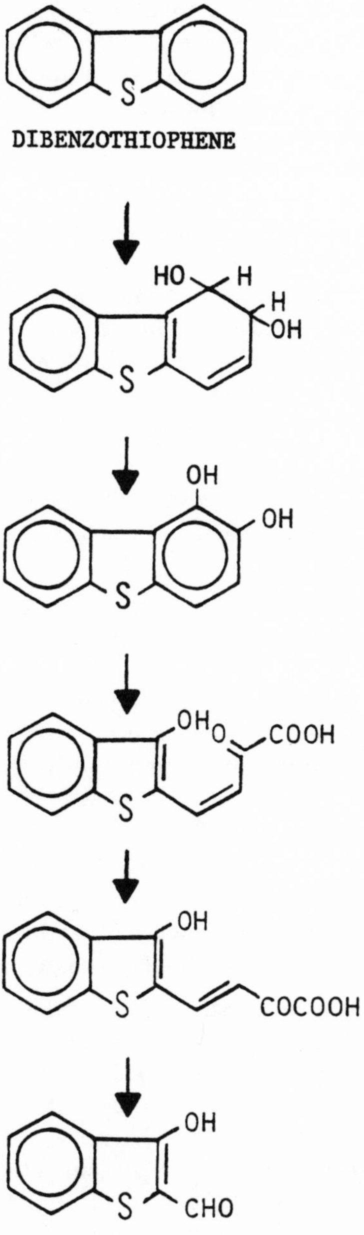

Figure 4.11: Pathway of microbial conversion of dibenzothiophene into water soluble compounds. (Source: Finnerty, W.R., 1980.)

organisms are used to concentrate metals from dilute waste streams. Common brewer's yeast, *Saccharomyces cerevisiae,* can accumulate uranium up to about 20% of its total weight. Settling ponds containing photosynthetic algae or rapidly growing aquatic vegetation, such as water hyacinths, are also fulfilling this purpose.

. The greatest R&D effort involving near-term applications of biotechnology to pollution control will be in developing improved microbial strains for decontamination of polluted waste waters and for *in situ* detoxification of contaminated soils and sediments. There exist considerable gaps in our basic knowledge of the types of microorganisms capable of degrading toxic chemicals. In particular, anaerobic bacteria and filamentous fungi represent two diverse classes of microbe for which considerable potential exists for biological pollution control, but little is known of their general properties.

Bacteria are classified into several groups based on the effect that oxygen has on their growth and metabolism:

> Obligate aerobes require oxygen for growth. An example is the tubercle bacillus, the causative agent of tuberculosis.
>
> Obligate anaerobes survive only in the absence of oxygen. Examples include clostridia (various species of which cause botulism, tetanus, and gangrene), bacteroids (intestinal bacteria that ferment glucose to form organic acids; e.g., formic, acetic, propionic, butyric, lactic, and succinic), denitrifiers that reduce nitrate to nitrogen gas, sulfate reducers that produce hydrogen disulfide (a source of pollution in anoxic ponds and streams), and methane producers that form marsh gas.
>
> Facultative organisms, such as many enteric bacteria (e.g., *E. coli*), can thrive with or without oxygen by shifting to different metabolic processes in each case.

Anaerobic bacteria are particularly relevant to pollution control practices because of their prevalance in subsoil. Thus, bacteria of this type will encounter toxic chemicals or petroleum wastes that have been spilled, as well as herbicides and insecticides that have been applied to the ground. A subgroup of anaerobic bacteria, called microaerophilic, can tolerate or even prefer low oxygen pressures (but much less than in air). These conditions prevail just beneath the surface of the soil. Thus, microaerophiles, about which very little basic information is known, should receive considerable attention for possible future use as *in situ* decontaminating agents. Likewise, anaerobic bacteria that thrive in underwater sediments, such as anoxic settling ponds or in the bottom of the kepone-laden James River, will be the subjects of more intense research in the years ahead.

Fungi are classified into three groups: (1) single-celled yeasts, (2) multicellular filamentous colonies, or molds, and (3) muchrooms. The filamentous fungi include some well known types, such as *Neurospora, Penicillium,* and *Aspergillus,* as well as lesser known aquatic water-molds and soil fungi. The genetics and biochemistry of fungi are much less well understood than are bacteria. However, it is certain that, like bacteria, fungi serve crucial roles in recycling organic matter throughout the biosphere. The contribution made by fungi to the decontamination of polluted soils and streams is becoming better appreciated, and research into the application of fungi to waste management should receive greater attention in the years ahead.

The following list outlines some aspects of applied genetics and waste management that will be under development.

> Cataloging the types of chemical transformations performed by microorganisms and the microbes involved.
>
> Isolating and characterizing the genetic material and enzymes responsible for the observed transforming activity.
>
> Conducting genetic engineering on organisms that occur naturally in a particular environment (e.g., river bed sediment) to confer the ability to degrade a pollutant that is not normally present in that environment (e.g., kepone). Successful decontamination of polluted sites by *in situ* biotreatment requires that the engineered microbe will compete favorably with existing microflora.
>
> Developing biotreatment systems for dealing *in situ* with specific wastes under a given set of conditions. For example, a chemical spill at a particular site may require a different microbe depending on the ambient temperature, or on the presence of certain nutrients. Exogenous nutrients such as glucose may have to be supplied.
>
> Designing bioreactors for on-line waste stream treatment. Systems for immobilizing microbes are under development. Monitoring and controlling the concentration of toxic substances in the waste stream are vital since excessive doses of most pollutants are deadly even to microbes that thrive on low concentrations of these chemicals. Thus, the design and engineering of

systems for diluting concentrated wastes prior to bio-treatment may be a greater technical challenge than is the development of microbial populations capable of performing the biodegradation.

5

Agricultural Applications and Trends

In this chapter, the scope of impacts of agricultural genetic engineering on the environment and public health is defined and discussed. Physiological and biochemical background information, as well as a discussion of environmental hazards and benefits, is provided for molecular vectors and genes of potential agronomic importance (Table 5.1).

Table 5.1: Genes and Gene Functions Whose Ecological and Social-Economic Impacts were Assessed*

Genes	Source of Genes	Potential Significance
Gene transfer vectors	*Agrobacterium tumefaciens*/plant DNA viruses	Insertion of new genes into crop plants
Nitrogen fixation *nif*	*Klebsiella/Rhizobium*	Genetic engineering of new nitrogen-fixing plants and microbes
hup	*Alcaligenes/Rhizobium*	Enhancing energy efficiency of symbiotic nitrogen fixation in legumes
lit	Blue-green algae	Key genes for production of reductant during photosynthesis
Denitrification *den*	*Klebsiella/Pseudomonas*	Important in soil fertility and water quality
Physiological stress *osm*	*E. coli/Salmonella*/higher plants	Osmoregulatory genes provide tolerance for drought, salt, and thermal stresses in microbes and plants
Photosynthesis *cfx*	*Alcaligenes*/higher plants	Enhancing efficiency of photosynthetic carbon dioxide fixation in plants

*The nif genes are clustered in an operon and are conveniently referred to as genes. However, the other genes listed; hup, lit, den, osm, and cfx; are all only short names for gene functions. They *may* involve many different genes which are not tightly linked.

AGROBACTERIUM (Ti PLASMIDS) AND PLANT VIRUSES– GENE VECTORS FOR AGRICULTURAL GENETIC ENGINEERING

Several vectors for transferring foreign genes into high plant cells or tissues are currently under development. These include *Agrobacterium tumefaciens,* caulimoviruses (e.g., cauliflower mosaic virus), a gemini-virus (e.g., bean golden mosaic virus), potato leafroll virus, and liposomes.

Below we assess the current data on *Agrobacterium* and plant DNA viruses as they relate to transfer of genes to plants and to hazards or benefits derived from dispersion of new genomes in the environment. Since liposomes are artifically constructed membranous vesicles, they cannot grow or divide, and therefore, they do not pose a threat to the environment or to human health.

Agrobacteria

Agrobacterium tumefaciens is a gram-negative, soil bacterium that causes a neoplastic disease of dicotyledonous plants called crown gall (Braun and White, 1943). Crown gall tumors can proliferate autonomously in tissue culture and, in a few cases, tumorous plants can be regenerated from the cells (Braun and Wood, 1976). The tumorous state of crown gall tissue results from the transfer of DNA from *A. tumefaciens* to the plant cell (Schell et al., 1979). This natural system for DNA exchange is being studied extensively. In addition to providing an interesting system of genetic exchange and tumor induction, *Agrobacterium* may make available a vector for the introduction of new genes into plants by molecular genetic engineering methods.

Both physical and genetic evidence indicates that the agents responsible for crown gall tumor induction are large plasmids, pTi (tumor-inducing plasmids), contained in all oncogenic strains of *A. tumefaciens* (Zaenen et al., 1974). Loss of these large plasmids results in loss of bacterial oncogenicity, and introduction of a Ti plasmid into originally non-oncogenic, plasmid-free *Agrobacterium* strains render them oncogenic.

The Ti plasmids of tumor-inducing strains of *A. tumefaciens* are large, ranging in size from 150 to more than 200 kilobase pairs (Depicker et al., 1978). These plasmids encode functions for oncogenicity, opine (arginine derivatives) biosynthesis and catabolism, conjugative transfer between bacteria, sensitivity to antibiotics, and exclusion of bacteriophage Apl (Holsters et al., 1980). Within the Ti plasmids is a special DNA segment called TDNA, which can stably integrate into the nuclear genome of transformed plant cells (Dell-Chilton et al., 1977; Yadav et al., 1980).

Compared to *Agrobacterium,* the plant viruses are less well developed as vectors for recombinant DNA experimentation. Information is lacking on essential virus-coded protein products and their locations on the genome. So insertions of foreign DNA into a viral genome may destroy the ability of the foreign gene to be expressed or to be replicated. Certain plant DNA viruses may be of limited use as vectors. For instance, cauliflower mosaic virus is not transmitted through seed. So only dicotyledonous plants that are not sexually propogated can be subjected to genetic transformation. Also, cauliflower mosaic virus probably requires a helper virus in order to replicate. New genes inserted into the cauliflower mosaic virus genome recombine with the helper virus genome thereby making transduction with a new gene very difficult (R. Meagher, personal communication).

Strategies for Introducing New Genes into Crop Plants

Two basic approaches are being developed to engineer the genes of plants (Figure 5.1). These center on utilizing plant DNA viruses or *Agrobacterium* (the Ti plasmid of *agrobacterium*) as vectors (Kado, 1979; Schell et al., 1979).

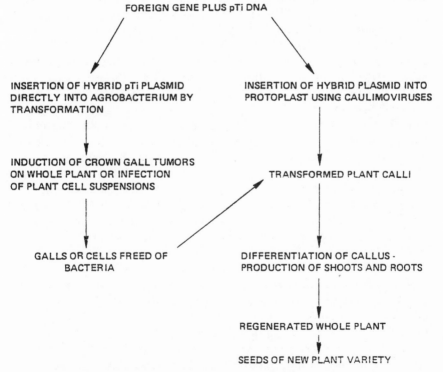

Figure 5.1: Flow diagram of two strategies for introducing foreign genes into higher plants.

Agrobacteria, and possibly other bacteria, have evolved this intricate plasmid mechanism to transfer specific genetic information to plants in such a way that the transformed plant cells express a number of new phenotypes, such as uncontrolled proliferation and the synthesis and probable release of opines. The bacteria, but not the plant, benefit from this transformation because they can selectively utilize the opines for their own growth and proliferation. The surface of crown galls therefore provide an ecological niche in which those bacteria that harbor plasmids with genes which enable the use of opines for energy, carbon, and/or nitrogen source have an important selective advantage over other soil bacteria.

Plant DNA Viruses

The simple genomes of specific plant DNA viruses may have some advantages not offered by other types of genetic vectors. Of the several hundred known plant viruses, only two or three have DNA as the genomic material. The others are RNA viruses. The plant DNA viruses are the caulimoviruses, the teminiviruses, and the potato leafroll virus. Little is known about the potato leafroll virus because it is present in only very small quantities in infected plants. Therefore, it will not be discussed further here. The caulimoviruses encompass at least six individual viruses with doublestranded DNA: cauliflower mosaic virus, carnation etched ring virus, dahlia mosaic virus, mirabilis mosaic virus, figwort mosaic virus, and strawberry vein banding virus (Shepherd, 1979). Of these, cauliflower mosaic virus is the most thoroughly studied (even though the data base is still relatively small) and is being seriously examined as a cloning vector for transferring genetic information to plants (Szeto et al., 1977; Lebeurier et al., 1980). Cauliflower mosaic virus infects species of the Cruciferae (e.g., cabbage, cauliflower, mustard), and some strains infect members of the Solanaceae (e.g., tobacco and carrot) (Broadbent 1957, Lung and Pirone 1972). The genome of cauliflower mosaic virus is a double-stranded, circular DNA molecule of about 7 kilobase pairs (Shepherd et al., 1970). A physical map of the cauliflower mosaic virus genome based on cleavage by restriction endonuclease has been determined (Meagher et al., 1977), and the viral genome can be clonally replicated in *E. coli* (Szeto et al., 1977).

Geminiviruses are a group of single-stranded DNA viruses including bean golden mosaic virus, beet curly top virus, cassava latent virus, *Euphorbia* mosaic virus, maize streak virus, and tobacco leaf curl virus. Of these, bean golden mosaic virus is the most studied. This virus infects members of the Leguminosae, such as *Phaseolus vulgaris* and *P. lunatus*. As compared with the caulimoviruses, geminiviruses have smaller and less complex particles with single-stranded DNA genomes that may be multipartite in nature.

In the first approach, useful new genes may be joined *in vitro* to the Ti plasmid, and the resulting hybrid plasmid can then be inserted by transformation into a plasmidless strain of *A. tumefaciens*. The transformed bacterium inserts the hybrid plasmid (or part of it) into the plant cell during infection of a whole plant. The infected plant tissues (crown gall) are excised from the plant and are brought into tissue culture. These tissue cultures are then encouraged to differentiate into whole, seed-bearing plants.

In the second approach, individual plant cells or protoplasts are transduced by direct insertion of a hybrid plasmid using liposomes or by employing DNA plant viruses, containing new genes. The protoplasts, now housing the new genes, can then be regenerated into whole plants that will produce seed.

Although both approaches are logical, they are clearly over-simplifications and many problems remain. Many genetic and biochemical barriers in plants (some yet to be identified) must be surmounted before genetic engineering can be successful (Kado, 1980). Even though the techniques for maintaining crown gall tissues in axenic culture are well developed, the regeneration of the tumorous tissues into whole plants is less well established. In fact, knowledge about the regeneration process has been a major limiting factor in developing most normal plant cell or protoplast cultures (Thorpe, 1978). Also, pieces of Ti plasmids found in plants derived from tumorous tissues are lost during seed production (Yang et al., 1980) so that the feasibility of generating seed stock of genetically modified crops is greatly reduced. This could mean that the use of Ti plasmids for genetic engineering of plants will have to be limited to plant species that can be reproduced by vegetative propagation. Another limitation is posed by the fact that *A. tumefaciens* and most of the plant DNA viruses rarely infect monocotyledonous plants. Many of our most important crop plants, including corn, wheat, and rice, are monocotyledons. However, many vegetable crops can potentially be engineered by *Agrobacterium* and plant DNA viruses.

Potential Environmental Benefits and Hazards

Agrobacteria: Several features of *Agrobacterium* could promote the spread of pTi DNA throughout the environment. The Ti plasmids are conjugative in that they carry genes that enhance conjugation and transfer of Ti plasmids between donor and acceptor bacteria (Kerr et al., 1977). Observations suggest that a conjugative mechanism might be involved also in the transfer of the Ti plasmid from bacteria to plant cells (Lippincott and Lippincott, 1975). The host range of Ti plasmids has been further extended by production of hybrid plasmids constructed by combining the Ti plasmid with the wide-range plasmid RP4. The Ti-RP4 plasmid can be transferred by conjugation at a high frequency to many other gram-negative bacteria, such as *Pseudomonas, Rhizobium,* and *Escherichia* (Schell and Van Montagu, 1978).

The promiscuity of Ti plasmids may be considered both an advantage and a risk. These plasmids offer chances to study the natural mechanism of DNA transfer among different organisms such as bacteria and plants, to examine the regulatory mechanisms involved in bacterial conjugation, and to understand the molecular biology of neoplastic transformation. On the other hand, plasmids that confer increased capabilities to *Agrobacterium* may be more easily passed to other Agrobacteria species (and other gram-negative bacteria). For instance, plasmid transfer may produce more virulent strains of crown gall, hairy root (White and Nester, 1980), and other plant diseases.

Given the risks, containment levels for experiments involving recombinant DNA which increases the virulence and host range of a plant pathogen beyond that which occurs in natural genetic exchange are determined by NIH on a case-by-case basis. In April, 1980. NIH ruled specifically on the use of *A. tumefaciens* as a host-vector system. According to that ruling, cloned DNA fragments from a nonprohibited source can be transferred into *A. tumefaciens* (containing a Ti plasmid) using a nonconjugative *E. coli* plasmid vector coupled to a fragment of the Ti plasmid and/or the origin of DNA replication of an *Agrobacterium* plasmid. Such gene transfers can be conducted under containment conditions one step higher then would be required for the desired DNA in HV1 systems. Transfer into plants or cells in culture would also be permitted at this higher containment level.

Several test decisions involving the NIH ruling about the genetic engineering of plant pathogens (e.g., *Agrobacterium*) have been made. For instance, it was agreed that a hybrid plasmid composed of (1) *E. coli* plasmid pB325, (2) the origin of replication and transfer genes of *A. tumefaciens* plasmid Ti (3) the thiamine gene of *E. coli,* and (4) *Arabidopsis* DNA, may be transformed into *A. tumefaciens* under P1 conditions. *A. tumefaciens* may subsequently be used to introduce the composite plasmid carrying *Arabidopsis* DNA and the *E. coli* thiamine gene into *Arabidopsis* plants under P1 containment conditions. Research with this hybrid plasmid is to be carried out by Dr. Donald Merlo at the University of Missouri, Columbia. In another case decision, the use of *A. tumefaciens* with the hybrid plasmid Ti-RP4 was considered as a HV1 system under P3 containment. Finally, in another instance, NIH has approved experiments involved in testing mutants of *A. tumefaciens* on carrot slices. The DNA of *Agrobacterium* and EK2 vector may be cloned back into *Agrobacterium,* and these transformed bacteria may then be introduced into plant cells in culture under P1 conditions.

NIH has also ruled on using nonpathogenic vectors for transferring DNA to higher plants. DNA from any nonprohibited source, that has been cloned and propagated in *E. coli* or *S. cerevisiae* vector can be used for cloning to any higher plant (angiosperms and gymnosperms) and propagated under conditions of physical containment comparable to P1 and appropriate to the organism under study. Intact plants or propagative plant parts may be grown under P1 conditions. Containment must be modified to prevent the spread of pollen, seeds, and other propagules. This can be accomplished by conversion to negative pressure in the growth cabinet or greenhouse or by the bagging of reproductive structures. Transfers to any other plant will be considered on a case-by-case basis.

Plant DNA Viruses: Several features of cauliflower mosaic virus make it an excellent candidate for a genetic vector but raise concern about environmental impacts of spreading recombinant rival genomes. The virus is transmissible mechanically to easily propagated herbaceous plants, its free DNA genome (i.e., when separated from the viroid) is infectious (Shepherd et al., 1970), and it has no tissue restriction. In addition, cauliflower mosaic virus is spread by insects and by infested, vegetatively propagated, planting material. Sucking insects, called aphids, are its only natural vectors. Feeding aphids pick up the viruses on their mouthparts and carry them to healthy plants during later feeding. Some aphid species can transmit viruses for several hours and occasionally for 1 or 2 days after acquisition. Aphids are common in greenhouses where unintentional transmissions may occur frequently. This problem can be prevented by use of non-aphid-transmitted strains of cauliflower mosaic virus. Transmission of this virus has been thoroughly reviewed (Shepherd, 1976).

One feature of geminiviruses that makes them good candidates for genetic vectors is their wide host range. Geminiviruses infect important crop plants in the Leguminosae and Gramineae families. The beet curly top virus, for example, can infect 300 species. However, this wide host range also increases the chances of dispersion of undesirable genes. In addition, whiteflies and leafhoppers, that transmit geminiviruses in nature, can maintain the viruses for the lifetime of the insect. The insects may acquire the virus during feeding periods of less than one hour. A brief latent period of 4 to 8 hours is required for the virus to circulate throughout the insect's body. The virus is absorbed through the gut wall into the hemolymph of the insect and is secreted by the salivary glands during feeding. Both whiteflies and leafhoppers are common in greenhouses, so unintentional transmissions may occur (Shepard, 1979).

Plant DNA viruses have several biological characteristics that could make them relatively safe genetic vectors. Cauliflower mosaic virus has a restricted host range (infects plants of the Cruciferae family) and will not infect such important grain crops as corn and wheat. Although the geminiviruses have a wide host range, they are restricted to one tissue, the phloem, and are not distributed throughout all somatic tissues. Only one geminivirus, the bean golden mosaic virus, can be mechanically transmitted to plants. Cauliflower mosaic virus can be mechanically transmitted through abrasion of leaf or stem surfaces. However, these DNA viruses are not transmissible by seed or pollen, and for this reason, accidental transmission from spillage of purified virus preparations is unlikely. Also, the fragility of plant protoplasts (into which the virus would be introduced), combined with several properties of the virus, provides adequate safety. Since no apparent risk to the environment from the use of a plant virus-protoplast system is envisioned, NIH has suggested that no special containment is necessary.

When plant DNA viruses are used as vectors in intact plants, or in propagated plant parts, the plant is grown under P1 conditions (i.e., in either a limited access greenhouse or a plant growth cabinet that is insect restrictive, that preferably is equipped with positive air pressure, and in which an insect fumigation regime is maintained). Soil, plant post, and unwanted infected materials must be removed from the greenhouse or cabinet in sealed, insect-proof containers and then sterilized. It is not necessary to sterilize run-off water from the infected plants, as this is not a likely route for secondary infection. When the viruses are used as vectors in tissue cultures or in small plants in axenic cultures, no special containment is necessay.

Infected plant materials that have to be removed from the greenhouse or cabinet for further research must be maintained under insect-restrictive conditions, which provide sufficient containment and which are similar to those required for licensed handling of exotic plant viruses in many countries.

NITROGEN FIXATION (nif) GENES

Background

The biochemical genetics of nitrogen fixation has been recently reviewed by Brill (1980) and Shanmugam et al. (1978). Nitrogen supply is probably one of the most important factors limiting plant growth (NSF, 1980). Even though nitrogen is an abundant atmospheric gas, only certain bacteria and blue-green algae possess the enzyme nitrogenase and are capable of fixing and reducing nitrogen to ammonia. Free-living nitrogen-fixing bacteria (i.e., those that are not intimately associated with a specific plant include members of *Klebsiella, Azotobacter, Clostridium, Rhodospirillum, Aspirillum,* and various cyanobacteria (blue-green algae). Bacteria that normally fix nitrogen only when they are symbiotically associated with a plant are *Rhizobium spp.* (which nodulate legumes) (Vincent, 1974; Fred et al., 1932), certain actinomycetes (which nodulate *Comptonia* and alder) (Torrey, 1978), and *Anabaena azollae* (a cyanobacterium that fixes nitrogen within the leaf pores of the water fern *Azolla*) (Peters, 1978). No eukaryotic organisms have been shown to fix nitrogen directly.

The biochemistry of nitrogen fixation is well understood. Nitrogenase is made up of two soluble proteins, called components I and II (Bulen and LeComte, 1966; Eady et al., 1972; Mortenson et al., 1967; Shah and Brill, 1973; Vandercasteele and Burris, 1970). Component I is also known as molybdenum-iron protein or nitrogenase while component II is known as iron protein or nitrogenase reductase (Hagaeman and Burris, 1978). Substrate binding and reduction take place on component I (Orme-Johnson et al., 1972; Smith et al., 1972; Smith et al., 1973). The role of component II is to supply electrons, one at a time, to component I (Ljones and Burris, 1978; Mortenson et al., 1967).

Since both components are rapidly and irreversibly inactivated by brief exposure to oxygen (Bulen and LeComte, 1966), organisms that fix nitrogen must have different mechanisms to protect their nitrogenases from oxygen. For instance, several nitrogen-fixing bacteria are strict anaerobes. Others, such as *Azotobacter,* have very high respiratory activity which rapidly lowers the internal oxygen concentration (Phillips and Johnson, 1961). Still others, such as cyanobacteria, package nitrogenase in heterocysts, specialized thickwalled cells that lack the oxygen-producing reactions of photosynthesis (Fleming and Hazelkorn, 1973; Peterson and Wolk, 1978, Tel-Or and Steward, 1977). Symbiotic nitrogen fixers such as *Rhizobium* protect their nitrogenase from oxygen inhibition by complexing it with the red pigment leghemoglobin, a myoglobin-like protein which can bind oxygen reversibly and with an extremely high affinity. Leghemoglobin is a product of a symbiotic interaction, the globin being produced by the plant, the heme by the bacteria (Beringer et al., 1979). Biological nitrogen fixation, like the industrial process used to produce nitrogen fertilizer (Haber-Bosch process), uses considerable energy. Between 12 and 36 molcules of adenosine triphosphate (ATP) are required for each molecule of nitrogen fixed (Brill, 1979).

The location of the nitrogen fixing (nif) genes has been determined in *Klebsiella pneumoniae.* The nif operon includes structural genes for the production of the nitrogenase subunits, regulatory genes, and genes for the synthesis of the molybdenum cofactor. Situated in a cluster near the his operon, the nif genes occupy less than 1% of the total bacterial chromsome. At least 17 nif genes make up the cluster, and no non-nif genes interrupt it (MacNeil et al., 1978). Complementation analysis with polar mutations (Elmerich et al., 1978; MacNeil et al., 1978; Merrick et al., 1978) and studies of fusions of lac with each gene (MacNeil and Brill, in press) indicate that the genes are organized into seven distinct operons and the transcription of the operons is in the same direction, namely, toward the his genes (Figure 5.2).

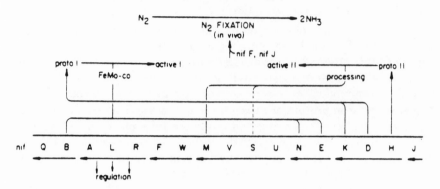

Figure 5.2: Order, transcriptional organization, and function of the *K. pneumoniae* nif genes. The arrows along the bottom represent the seven individual operons (Source: Brill, 1980).

The nif region is approximately 24 kilobases long (Canno et al., 1979). A physical map of a large part of the nif region has been produced by restriction endonuclease cleavage analysis of cloned DNA (Reidel et al., 1979). The protein structure of nitrogenase component II is encoded by nif H, but the product of nif M, nif S, and possibly nif V genes are necessary to modify and activate component II (Roberts et al., 1978). Two proteins in component I are specified by nif K and nif D. In order to obtain active component I, an iron molybdenum cofactor (FeMo-co) is required. FeMo-co synthesis requires the products of nif Q, nif B, nif N, and nif E (Roberts et al., 1978). Proteins coded by nif F and nif J genes are necessary to transport electrons to nitrogenase. The electrons break the triple bond between the two nitrogen atoms, converting each atom to a moleucle of ammonia (Nieva Gomez et al., 1980; Robert et al., 1978; Yock, 1974; Yates, 1971). The activator produced by nif A turns on the production of the nif-coded proteins when the cells lack sufficient fixed nitrogen in their environment. Both oxygen and ammonia repress nitrogen fixation. The function of the nif L product is to prevent nitrogenase synthesis when the cells start aerobic functions. The site between nif L and the promoter for the nif RLA operon at which ammonia regulation occurs is defined as nif R. It is not yet known whether nif R codes for a protein (Roberts et al., 1978; MacNeil and Brill, 1978; St. John et al., 1975). Nif W and nif U are also not essential under normal conditions (Roberts et al., 1978). In *Rhizobium,* some genes for nitrogenase, as well as for many other genes involving symbiosis and nitrogen fixation, are localized on a plasmid. This was shown recently by hybridization of plasmid DNA with the cloned structural nif genes of *Klebsiella penumoniae* (Krol et al., 1980).

The nif genes have so far been transferred by the techniques of DNA recombination only among microbes. The genes for *Rhizobium trifolii* have been transferred to *Klebsiella aerogenes,* and preliminary reports of nif gene transfer in photosynthetic microbes have appeared. *E. coli, S. typhimurium,* and *K. aerogenes* all accept *K. pneumoniae* nif genes. Other organisms, such as *Agrobacterium tumefaciens, Rhizobium meliloti, Erwinia herbicola,* and *Azotobacter vinelandii,* accept nif plasmids from *K. pneumoniae.* However, the new recombinant organisms do not fix nitrogen. Since some of the recipient bacteria do not express the nif genes completely (even if they accept them) while others do not accept them at all, more genes than just the nif cluster obviously are needed for nitrogen fixation, even in bacterial hosts. For example, genes that control the level of free oxygen in the cell are necessary to prevent nitrogenase from being destroyed (Postgate, 1977). Also, genes that regulate the production and activity of nitrogenase and other accessory proteins are required.

The nif gene cluster may have been transferred to plants in nature (Postgate, 1977). However, the simple transfer of nif genes does not confer the ability to fix nitrogen. Nitrogen-fixing bacteria also have the mechanisms for regulating, transcribing, and translating the information on nif and its supporting genes. The means for using this bacterial information will thus have to be introduced along with the nif cluster and whatever else is transferred to plants. The existence of this genetic barrier between prokaryotes and higher plants will make the transfer of nif genes to plants a long-term project (1 to 2 decades). There are several possible mechanisms for transferring this information:

> Plant DNA viruses have sites that can be read by a plant's genetic machinery so these may be useful vectors (see discussion on vectors earlier in this book).

> Some plant organelles (mitochondria and chloroplasts) have more bacterium-like genetic equipment than the main plant chromosomes and, therefore, these organelles might be more satisfactory hosts. It should be noted, however, that mitochondrial codons differ somewhat from the codons used by all other organisms. Other problems with organelles as gene vectors include multiple genome copy number, multiple organelles within a cell, sorting out, and organellar recombination.

> Deliberate construction of more effective symbiots, such as endotrophic mycorrhizae, might be used to

effect the transfer of information. Ecological studies
are now uncovering more such associations existing
naturally in many environments (Postgate, 1977).

Even if nif genes cannot be directly transferred to crop plants, other
important advances in enhancing fixation using recombinant DNA
techniques may be possible. For example, the regulation of nif can be
altered by mutation, so that such free-living nitrogen fixers as *A. vinelandii* or *K. pneumoniae* fix nitrogen even when they are supplied with
ammonia. These mutants would be depressed. Partially derepressed
mutants might also be obtained that would yield more nitrogen to their
hosts (Gordon and Brill, 1972; Gordon et al., 1975). In addition nitrogen-
fixing microbes with only one set of nif genes might be given a second
copy of nif genes, which might make them more efficient nitrogen fixers.
Another possibility is to increase the length of time that bacteria associated with plants fix nitrogen since these bacteria often stop fixing nitrogen during seed formation. A few more days fixation, which might
dramatically increase the protein content of crops, could be induced by
genetic manipulation (Postgate, 1977).

Potential Environmental Benefits and Hazards

Modern farming methods have reduced the amount of organic matter in soils to an equilibrium value of 40-60% of the original amount.
Soil fertility (inorganic matter) has deteriorated as well. Crop yields
can only be maintained through the continued application of energy-
intensive, chemical fertilizers.

Crop productivity (yield per unit land area) during the last quarter
century is highly correlated with the input of nitrogen fertilizer. For example, the yields of cereal grains in both industrialized and less developed countries from 1950 to 1975 paralleled the increasing rates at
which fertilizer has been applied (Hardy et al., 1975). Up to one-half of
the 3% average annual increase in world grain production was due to
the exponential increase in the use of nitrogen fertilizer. U.S. fertilizer
use during this 25 year period increased by a factor of 2.7, and nitrogen
use increased by a factor of approximately 10 (Paul et al., 1977).

From data supplied by the USDA (USDA, 1980), projections of nitrogen fertilizer use in 1990 have been made. On a worldwide basis, 1.4
times as much fertilizer will be consumed in 1990 as in the mid-1970s.
Usage in the developing countries is expected to increase by a factor of
1.7, in the developed countries by 1.2 and in the centrally planned countries by 1.6.

Environmental effects of this increased use of fertilizer include
greater depletion of the ozone layer by the nitrous oxide in fertilizer, increased eutrophication in aquatic environments by leaching of applied
fertilizer into watercourses, lakes, and reservoirs, and the contamina-

tion of drinking water supplies with nitrates, which could increase the incidence of methemoglobinemia, a disease which affects primarily infants under the age of three.

The enhancement of biological nitrogen fixation by recombinant DNA techniques could result in increased crop productivity without increased use of fertilizer and without the detrimental effects noted above. In additon, energy and dollar savings in industrial nitrogen production (Haber-Bosch process) and transportation would occur. Approximately $10 billion are spent annually on chemical fertilizers in the U.S. alone (*Chemical Week,* 1980). If nitrogen were not a limiting factor, other plant nutrients such as potassium, sulfur, magnesium and phosphorus would become limiting to plant growth. If, by proper husbandry, these were attended to then the rate of carbon dioxide fixation would become limiting, as it is today with otherwise well-fertilized legumes. Techniques of genetic engineering may provide solutions to this limitation (also see section below on carboxylase genes) (Postgate, 1977).

Detrimental effects of enhancing nitrogen fixation using recombinant DNA techniques can be envisioned, but they are slight compared to the potential benefits. Any plant or microbial system that fixes nitrogen must consume more ATP than it does when ammonia fertilizers are used. No fewer than 7 to 8 ATP molecules would be required per molecule of ammonia received by the plant, and the number could be greater if a large amount of photosynthate had to be expended (and therefore ATP lost) to protect nitrogenase from oxygen. Under identical environmental conditions, a nitrogen-fixing cereal would probably not be as productive as one supplied with comparable amounts of ammonia. One consequence of this increased energy need might be that the range of a crop might move towards the equator, but by what distance is difficult to judge. In any event, farmers might find the lower yields from nitrogen-fixing crops acceptable to the extent that reduced income was more than compensated for by lower imputs (and costs) of nitrogen fertilizer.

If the property of nitrogen fixation became more widespread in wild plants, more vigorous growth of weeds and natural grasses might become a problem. This seems unlikely for two reasons. First, most common crop plants are self-pollinated. Crossing with related weed species occurs less than 1% of the time (D. Ray, personal communication). Second, growth of nitrogen-fixing weeds would only be more vigorous in seriously nitrogen-deficient soils, an unlikely condition under modern agricultural practice.

Transfer of nitrogen fixation to plant pathogens is another potentially harmful possibility. For example, if the plant-rotting microbe *Erwinia* or its host plants developed nitrogen fixing ability, pathogenicity would be increased because plants low in nitrogen are usually more resistant to *Erwinia* than plants with high levels of nitrogen

(Postgate, 1977). However, the major factor that usually influences plant susceptibility to *Erwinia* attack is moisture, not nutrition (D. Coplin, personal communication).

As nitrogen is often a limiting nutrient for decomposition of soil organic matter, increased nitrogen fixation by plants could eventually lead to reduction of soil organic matter both degrading soils and releasing more carbon dioxide into the atmosphere (Barney, 1980).

One can speculate on the impact of transfer of nitrogen fixation to the *E. coli* of a ruminant animal. This probably would have no effect because the rumen contains repressive amounts of ammonia, and the limited amount of air in the rumen would probably limit the quantities of nitrogen fixed. Derepressed nif (which is not repressed by ammonia) might yield nitrogen-fixing animals that could live on cellulose substrates without added protein (Postgate, 1974). It might also kill the animals through ammonia toxicity (Postgate, 1977). This possibility seems very remote since weakened strains of *E. coli* are used in recombinant DNA research. Ammonia toxicity caused by overfeeding of protein is more of a problem (R. Conrad, personal communication).

HYDROGEN UPTAKE (hup) GENES

Background

During nitrogen fixation, some of the electrons destined for nitrogenase do not combine with nitrogen gas to form ammonia but instead react with H^+ ions to form hydrogen gas. This process consumes ATP. Some bacteria have a mechanism to recover some of the electrons and thereby reduce the ATP lost by hydrogen evolution. These bacteria contain the enzyme hydrogenase which oxidizes hydrogen and yields either H^+ ions or electrons capable of reducing nitrogenase or energy in the form of ATP to drive nitrogen fixation (Figure 5.3) (Brill, 1979).

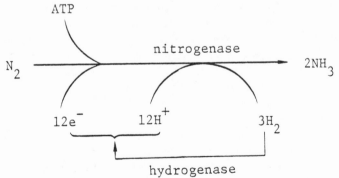

Figure 5.3: Nitrogen fixation and the hydrogen uptake mechanism. (Source: Brill, 1979.)

Several agronomically important species of *Rhizobium* do not have an active hydrogen uptake system (Schubert and Evans, 1976; Lim, 1978). In a recent study conducted at 70 different locations in the major areas of soybean production, about 75% of all *Rhizobium japonicum* strains tested lacked hup activity (Lim et al., 1980). One of the easiest ways to enhance nitrogen fixation by legumes would be to introduce hydrogen uptake (hup) genes for the enzyme hydrogenase or other genetic systems for hydrogen uptake mechanisms into the *Rhizobium* strains that lack them. The term hup genes is used simply to refer to genes which code for hydrogen uptake mechanisms. This process may be the product of a single gene or, more likely, multiple genes which may or may not be tightly linked.

Hydrogen bacteria, which use hydrogen as an energy source, may serve as a convenient source of hup genes. The hydrogen uptake mechanisms in these bacteria is essentially the same as that shown in Figure 5.3. Large plasmids that determine the ability to metabolize hydrogen in several species of hydrogen bacteria have been identified. For instance, the hup genes have been transferred from *Nocardia opaca* to *Nocardia erythropolis* by a natural conjugation system; the capacity was lost after exposure to mitomycin (Schubert and Evans, 1977). This suggests that the genes are situated on a transferable plasmid. Six other strains of bacteria (three strains of *Alcaligenes eutrophus, A. paradoxus, P. facilis,* and *P. Palleronii* also lost their ability to use hydrogen as an energy source after exposure to mitomycin. The authors concluded that the *A. eutrophus* strains, and probably the other species of hydrogen bacteria, contained a large plasmid required for hydrogen metabolism and autotrophic growth (Lim et al., 1980). Plasmids have been detected in several strains of *R. japonicum*. So far there is no correlation between presence of the plasmid and the hydrogen uptake ability in *R. japonicum*. Further characterizations and modifications of plasmids from hydrogen uptake bacteria may allow their introduction to other *Rhizobium* species (Lim et al., 1980).

Potential Environmental Benefits and Hazards

Data indicate that a hydrogen uptake system significantly benefits overall plant productivity, since energy supply is a major limiting factor in the fixation of nitrogen by soybeans (Hardy and Havelka, 1975). Hup$^+$ strains of *R. japonicum* are more efficient symbionts than hup$^-$ strains in comparisons of soybean productivity (Lim, 1978; Schubert and Evans, 1977). Also, the plants inoculated with hup$^+$ strains had higher yields than those inoculated with hup$^-$ strains (Albrecht et al., 1979).

At this time there seems to be no hazard in transferring hup to *Rhizobium* or other bacterial species that lack these genes since so many bacterial species already possess these genes.

WATER-SPLITTING (lit) GENES FOR GENERATION OF REDUCTANT

The symbiotic fixation of nitrogen is often limited by photosynthetic production by the green plant. Photosynthesis results in the production of both reducing power (NADPH ultimately derived from water) and ATP. The reaction involved is the Hill reaction:

$$\text{light} + H_2O + NADP \rightarrow \tfrac{1}{2}O_2 + NADPH + H^+$$

Blue-green algae (cyanobacteria) are the simplest organisms possessing water-splitting genes. Much basic research is required to isolate and characterize these genes before any recombinant DNA techniques can be applied (Andersen et al., 1980). Possibly the water-splitting genes could be transferred from blue-green algae to symbiotic nitrogen-fixing bacteria. The bacteria would still require carbohydrate from the plant, but the bacteria could provide some of its own reducing power.

DENITRIFICATION (den) GENES

Biological denitrification is the enzymatic conversion of nitrate and nitrite into gaseous forms of nitrogen, including nitrous oxide. Denitrification does not usually occur in well cultivated soils, but it does occur in poorly aerated soils containing an abundant supply of carbohydrates. During the process, oxygen is evolved. A variety of organisms, including *Klebsiella* and *Pseudomonas,* contain denitrification (den) genes, but as is the case with the lit genes, much basic research will have to be done before recombinant DNA techniques can be considered. A beneficial aspect of denitrification is that nitrogen pollution can be removed from rivers. Similarly, sewage waters containing high levels of soluble nitrogen can be purified through biological denitrification. One detrimental aspect is that under some circumstances denitrification may contribute to the production of nitrogen oxides (air pollutants found in smog). In addition, denitrification works against nitrogen fixation by diminishing available levels of soil nitrogen and requiring higher inputs of fertilizers (Andersen et al., 1980).

OSMOREGULATORY (osm) GENES

Background

The loss of increasing amounts of agriculturally productive land due to increased salinity is a serious problem worldwide. High salt concentrations in the soil create high osmotic pressures, thereby, reducing the availability of soil water to plants and causing plant cells to lose water.

Whole plant mechanisms for combatting excessive salt levels in the environment include osmotic barriers in the roots, swelling of leaves and other structures by absorbance of water which keeps the salt concentration constant and results in succulence, an exudation of salt on plant surfaces or by salt glands. Such mechanisms are under complex genetic control and are not easily amenable to genetic engineering. The study of halophytic plants and bacteria also indicates a cellular mechanism that may overcome excessive osmotic stress from the environment. The halophytic plants, which live and reproduce in or on oceans, seashores, estuaries, deltas, salt marshes, and saline desert soils, can survive by increasing the concentration of dissolved solutes within their cells, either by taking up salts or other solutes from the environment or by synthesizing soluble organic compounds (Epstein et al., 1980). The synthesized compounds include amino acids and their derivatives, such as proline, glycine, glutamate, or gamma-aminobutyrate; polyols (compounds with three or more hydroxyl groups), such as glycerol and sorbitol; and derivates of betaine, a substituted ammonium compound (Marx, 1979; Measures, 1975). Bacteria such as *Escherichia, Klebsiella,* and *Salmonella* probably use the same mechanisms to combat salt stress, as do plant cells.

A set of genes called osmoregulatory (osm) genes function in many plant and bacterial cells to control the mechanisms that prevent water loss (Flowers et al., 1977). Workers at the University of California, Davis, have been able to transfer plasmids that encode for salt tolerance and L-proline over-production in mutant strains of *Salmonella* to related bacteria. Mutant strains were selected by growing bacteria in the presence of normally toxic concentrations of the L-proline analog, L-azetidine-2-carboxylate. Mutants, that excrete L-proline, dilute and thus antagonize the analog. The mutation conferring L-azetidine-2-carboxylate resistance and salt tolerance is closely linked to the locus for the first enzyme of the L-proline biosynthetic pathway, proB, and it is transferred via an F' factor (a plasmid that is transmissible among the Enterobacteriaceae family). Researchers are still many years away from being able to isolate and fully characterize these genes (Marx, 1979; Andersen et al., 1980). However, advances should come more rapidly than might otherwise be expected since over 60 laboratories around the world are investigating the genetic aspects of plant resistance to damaging salt concentrations.

Plant and cell culture techniques, without recombinant DNA techniques, can also be used to obtain salt-tolerant plants (Rains, 1979; Croughen et al., 1978). This is discussed further in the section on mutation induction using plant tissue culture.

Potential Environmental Benefits and Hazards

Estimates of present saline soils on a worldwide basis range from 1 billion to 2.3 billion acres (Ponnamperuma, 1977; Massoud, 1974). Approximately one-third of the irrigated lands are now increasingly affected by salinity. The affected land amounts to 41 million acres in the U.S. and 198 million acres worldwide. This loss represents about 0.04% of the world's total irrigated land. If it remains constant through the year 2000, about 4.3 million acres (approximately 0.9%) of the world's total irrigated land will be out of production. These losses have a greater impact on food production than averaged figures suggest because irrigated land is almost always the most productive land in any region. Even assuming average productivity, 4.3 million acres represents the food supply for more than 5.75 million people (Barney, 1980).

In the U.S., salinity problems are becoming most critical in the highly irrigated southwest and west (Epstein et al., 1980). Irrigation waters often contain a higher soluble salt concentration than rainwater. Unless drainage is provided to carry off the irrigation waters, the water with its dissolved salts sinks deep into the soil. As the water table rises, the salt is carried back to the surface where it can inhibit plant growth and form a mineral crust. Soil amendments and fertilizers add to the problem. In the San Joaquin Valley of California alone, increasing salinity is costing farmers $32 million/year in reduced crop yields (Marx, 1979). About 400,000 acres of irrigated farmland are affected by high, brackish water tables that pose an increasingly serious threat to productivity. About 1.1 million acres–about 13% of the total valley–ultimately will become unproductive unless subsurface drainage systems are installed. The salting problems of the valley have been compared to those that resulted in the collapse of civilization in Mesopotamia and Egypt's upper Nile. Similar problems are also being observed in California's Imperial Valley (Barney, 1980).

Some nations, such as India and Australia, are largely in arid or semiarid zones. Thus most of their agriculture requires irrigation. As more land is put into production and the demands for water from agricultural, mining, recreational, domestic, and industrial interests increase, the problem of increased soil salinity grows.

Conventional breeding techniques currently are being used to adapt plants to saline conditions. An example of the saline culture research is that which has been performed at the University of California, Davis (Epstein, 1977; Epstein and Norlyn, 1977; Epstein et al., 1980; Rains, 1979). Highlights of the research include the following:

> Thousands of genotypes of barley and wheat have been screened.
>
> In tomatoes, exotic germplasm has been used to trans-

fer salt tolerance from an economically useless species to a commerical line.

Successful crop selections and new breeds have been field-tested under saline conditions (Epstein et al., 1980).

In view of this progress and because conventional breeding techniques pose no hazards to the environment, recombinant DNA techniques may not prove especially useful in the development of saline-tolerant crops.

The primary benefit of enhancing the salt tolerance of crops through breeding methods is the increased amount of land that can be brought under cultivation (with all of the accompanying economic and social benefits). Lands that are not in production because the soil is saline–naturally, because salt has accumulated under agricultural practice, or because the only available water is saline–could be cultivated with crop plants possessing increased salt tolerance. Crop production would likely be affected by increasing the amount of land under cultivation rather than increasing productivity as is the case with nitrogen fixation. In fact in the short term it is expected that yields of salt-tolerant crops grown under excessive salinity will be significantly lower than average world yields. However, small yields are better than no yield at all, something that would occur if currently cultivated lines were exposed to excessively saline conditions (Epstein et al., 1980; Barney, 1980).

Enhanced salt tolerance would also affect cropping patterns. However, economic forces (crop prices) as well as the amount of land under cultivation will continue to be the overriding influence on cropping patterns. Perhaps the secondary benefits of cultivating plants with enhanced salt tolerance include better control of soil erosion, maintenance of water table levels, and reduced flooding in some areas. Further geographic, economic, and social studies are needed to assess these effects (Barney, 1980).

Hazards associated with laboratory transfer of osm genes to plants seem small. However, it does seem remotely possible that salt-tolerant germplasm might be transferred in the field to weeds if crop plants with their new trait normally cross with related weed species (Postgate, 1977). Superior characteristics developed by conventional breeding techniques can and in fact now are being transferred to weed species in this manner. The transfer of traits (introgression) between sorghum and related wild types such as Johnson grass is an example. Not only are new traits transferred by introgression to the wild type, but wild types act as a gene pool for some characters of the cultivated forms. If a new cultivated sorghum is introduced, it will gradually acquire some characters of the former cultivated sorghum via introgression with the permanent wild types (Doggett, 1970). The magnitude of the weed problem is not likely to worsen, however. If the new trait allows an extension of a week's cultivated range or an increase in vigor, the range or vigor of the wild type weed will change proportionately.

CARBOXYLASE (cfx) GENES

Background

To meet the increasing demands of the world's growing population for food, agricultural yields will have to rise an average of over 2% a year over the next 25 years. This will approximately double primary food production according to a study of the National Academy of Sciences (1975). Carbon fixation is extremely important in crop productivity because 95% of the dry weight of plants is derived from photosynthesis (Bukovac et al., 1975).

Research on photosynthesis has revealed a number of possible ways to increase crop yields:

> Modification of photorespiration and carboxylation reaction
>
> Enhancement of photosynthetic electron transport
>
> Enhancement of translocation of photosynthetic products from leaves to other plant parts
>
> Enhancement of transport of carbon dioxide from the atmosphere to the chloroplast (Zelitch, 1979)

The mechanisms and genes involved in these processes have only begun to be elucidated. Also, the genes involved in the above processes are located in both the chloroplasts and the nucleus.

Photorespiration and Carboxylation

Photorespiration as carbon dioxide evolution differs biochemically from normal respiration in the dark and is specifically associated with the oxidation of compounds produced during photosynthesis. In tests on nine species, photorespiration was found to be 2.2 to 4.8 times as great as regular dark respiration (Decker, 1957). At high irradiance, release of carbon dioxide by this process in many C_3 plants is at least 50% of the net carbon dioxide fixed by photosynthesis. Photorespiration is undetectable or operates at only marginal rates in C_4 species. The rate of photorespiration is greatly dependent on the oxygen concentration at normal carbon dioxide levels, and the process is strongly inhibited in leaves when the leaf canopy is enriched with carbon dioxide (Jackson and Volk, 1970).

Some researchers have speculated that millions of years ago when the earth's atmosphere contained more carbon dioxide and less oxygen than now, C_3 plants would have flourished. However, as carbon dioxide was absorbed by plants performing photosynthesis and as oxygen levels increased, these species have been left with the evolutionary burden of photorespiration.

By blocking photorespiration, the net photosynthetic incorporation of carbon dioxide in many species might be increased by at least 50%. Although work is still to be done on the biochemistry of the process, evidence suggests that the oxygenase activity of ribulose bisphosphate (RuBP) carboxylase furnished phosphoglycollate, the initial substrate of photorespiration (Andrews and Lorimer, 1978) (Figure 5.4).

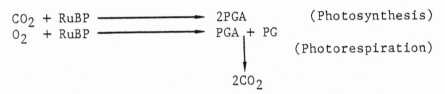

Figure 5.4: Carboxylase and oxygenase activities of ribulose bisphosphate carboxylase. RuBP = ribulose bisphosphate. PGA = phosphoglyceric acid. PG = phosphoglycollate (Source: Andersen et al., 1980).

Altering the RuBP carboxylase in such a way that oxidative cleavage of RuBP will be decreased or will not occur may reduce photorespiration. However, it is important to determine whether such an alteration is possible and productive before attempting to engineer the genes of higher plants. Studies on mutationally altered RuBP carboxylases that have a range of catalytic activities in strains of the bacterium, *Alcaligenes eutrophus,* are now underway (Andersen et al., 1980).

A second possible way of increasing carbon dioxide fixation or reducing photorespiration in some species is by transferring C_4 photosynthetic capability to C_3 plants. The biochemistry of the C_4 process is well understood, but the location and function of the genes are not known.

The process involves a complex compartmentation of carbon dioxide-fixing enzymes and carbon transport among cells. The C_4 process is a pump which increases carbon dioxide concentration in the chloroplast and reduces competition with oxygen (lowers photorespiration). Phosphoenol pyruvate carboxylase functions in the mesophyll cells inside leaves to produce a C_4 acid. This organic acid is then transported to the specialized bundle sheath cells that surround the conducting elements in the leaf, and there the acid is decarboxylated. Finally, RuBP carboxylase fixes the carbon dioxide released in the bundle sheath cells by

C_3 reactions of the Calvin cycle (Hatch and Slack, 1970; Leatsch, 1974).

A number of important crop plants (corn, sugarcane, sorghum, foxtail, and finger millets, and forage species (Bermuda grass, Dallis grass, Sudan grass and Rhodes grass), as well as certain weeds (crabgrass, pigweed, and others), are C_4 plants (Loomis et al., 1971). Leaves of C_4 species have high rates of net carbon dioxide assimilation (42-85 mg carbon dioxide per square decimeter per hour) and low rates of photorespiration at high levels of irradiance in air at 25° to 35°C. On the other hand, C_3 species assimilate carbon dioxide at rates about half those of C_4 plants and have rapid rates of photorespiration. The exceptions to this rule, such as sunflower and cattail, are C_3 species with rapid rates of photosynthesis. An understanding of the way these species overcome the rapid photorespiration exhibited by other C_3 plants would provide clues for improving photosynthetic efficiency (Zelitch, 1979).

Transferring the C_4 pathway into some Calvin cycle (C_3) species by conventional breeding of *Atriplex rosea* (C_4) and *Atriplex patula* (C_3) yields progeny that have various physiological and biochemical characteristics of photosynthesis intermediate between the parents. Several loci are involved, and transfer is difficult. Other genera known to contain species with both the C_3 Calvin cycle and the C_4 pathway are *Kochia, Bassia, Panicum, Euphorbia,* and *Cyperus.* However, interspecific crosses have not been attempted in these genera (Loomis et al., 1971).

Electron Transport

The rate of photosynthetic electron transport may control carbon dioxide fixation at saturating carbon dioxide levels. This has been shown in experiments with isolated chloroplasts. However, there is no evidence that this is also true for leaves in normal air. Electron transfer between water and $NADP^+$ in the light reactions of photosynthesis takes place along a chain of electron carriers. The carriers are organized into two systems, photosystem I and II, and are connected in series (Velthuys, 1980). The reaction chain of each photosystem contains several hundred chlorophyll molecules, in addition to accessory pigments, lipids, and proteins, that together functions as a photosynthetic unit. Light (400 to 700 nm) is collected by chlorophyll molecules, and the energy is transferred to the reaction centers of the photosynthetic where electron transfer begins. Some scientists assume that decreasing the size of the phtosynthetic units and increasing the number of reaction centers per chlorophyll molecule would increase photochemical efficiency, but this is yet to be proven. Although researchers are accumulating information on transcription-translation sites of the proteins involved in electron transport (Nasyrov, 1978), much work is still to be done. The genes that control or encode photosynthetic electron transport elements have not been identified or isolated.

Translocation of Photosynthetic Products

The rate at which photosynthetic products move from leaves to other plant organs may be an internal control of net photosynthesis. It is not currently known whether the rate of translocation is regulated by the loading of carbohydrate into the phloem, by the long distance movement of carbohydrate to its final destination, or by the unloading of carbohydrate at the point of storage or metabolism (Minchin and Troughton, 1980). Accumulation of certain metabolites in leaves may regulate photosynthesis and related processes by feedback-type mechanisms. In addition, translocation may have subtle influences by controlling the concentration of key metabolites that affect photosynthesis. It would be of great importance to be able to control the translocation process, thereby promoting photosynthesis, and also to be able to direct more photosynthate into the part of the plant to be harvested (Zelitch, 1979). However, much physiological research is still required to establish clearly the relationship between photosynthesis and levels of photosynthetic products (Loomis et al., 1971). The genetics of translocation have not been investigated. Applications of genetic engineering techniques are still many years away.

CO_2 Diffusion

There are several barriers to the diffusion of carbon dioxide from the atmosphere to the chloroplast. These barriers include the diffusion resistance imposed by the microscopic pores (stomata) on the leaf surface, the physical diffusion resistances caused by the aqueous medium of plant cells, and the diffusion resistances caused by carbon dioxide released by respiration and photorespiration. These barriers seem to be much more limiting to photosynthetic rates than photochemistry or enzymatic carboxylation. The effects of diffusion barriers may be reduced by optimizing the interception of solar radiation, by planting varieties whose leaves expand rapidly, by spacing plants to improve light interception, and by increasing the ratio of growing to senescent leaves (Loomis et al., 1979; Aelitch, 1971). Much physiological research is still required in this area, and applications of recombinant DNA techniques seem remote at this time.

Potential Environmental Benefits and Hazards

Since the recombinant DNA technology needed to improve the photosynthetic efficiency of crop plants may require years to develop, it is difficult to analyze its impact thoroughly. The benefits of higher productivity of wheat and other crop plants would be profound. For example, higher yields per unit of land area could be achieved without increasing the amount of fertilizer or pesticide used. This could result in better nutrition in less developed countries and lower prices with increased export trade in the developed countries.

A potential hazard, mentioned in the sections on nitrogen fixation and salt tolerance, is the transfer of enhanced photosynthetic qualities to weed species, a situation that might eventually require a greater use of herbicides.

INDUCTION OF MUTATIONS BY PLANT TISSUE CULTURE TECHNIQUES

Background

Such plant cell tissue culture techniques as the selection of somatic-cell variants, cell fusion, and the induction and selection of mutants, can create genetic variability and modify the genetics of plants, without resort to recombinant DNA techniques. Plant tissue culture techniques have been used to select cell lines tolerant to salinity, herbicides, and disease. Among the cell lines selected are:

> Tobacco and bell pepper lines tolerant to salinities as high as 1 to 2% (Dix and Street, 1975)
>
> Wheat (Rains, 1979) and alfalfa (Croughen et al., 1978) with enhanced salt tolerance
>
> Tobacco (Chaleff and Parsons, 1978; Radin and Carlson, 1978) and white clover (Oswald et al., 1977) lines selected for herbicide tolerance
>
> Sugarcane lines resistant to *Herminthosporium sacchari* sugarcane mosaic virus, and *Sclerosporia sacchari* (Nickell, 1977)
>
> Potato lines with increased frequency of genetic variation for agronomically valuable characteristics such as disease resistance (Shepard et al., 1980; Gwynne, 1980)
>
> Texas male-sterile cytoplasm maize lines resistant to *Helminthosporium* race T pathotoxin (Gegenbach et al., 1977)

However, while selection for cell lines tolerant to stress conditions has been accomplished, testing for genetic stability and tolerance at the whole plant level has rarely been attempted.

The above-mentioned species are all vegetable or grain crops. Tissue culture techniques can also be used to modify other plant species and to shorten the time required to develop them as new fuel or fiber crops. Many plants, which are commonly considered weeds because they compete with crops or other desirable plants, are now being considered as potential crops themselves (Buchanan, et al., 1978; Saterson et al., 1979). Among such nonconventional weed crops are:

> Common milkweed for fats, proteins, hydrocarbons, rubber, and bast fibers
>
> Giant ragweed for proteins, oils, rubber, paper pulp, and biomass for fuel production
>
> Jimsonweed for alkaloids as well as for biomass for fuel production
>
> *Euphorbia* (various species) for hydrocarbons and epoxy fatty acids

So far, applications of tissue culture to new crop development have been confined primarily to model systems. For example, tissue culture techniques have been applied to milkweed (Groet and Kidd, 1980) as well as to *Euphorbia* (Biesboer and Mahlberg, 1979). However, large-scale propagation for field planting or selection of stress-tolerant strains has not yet been attempted.

Potential Environmental Benefits and Hazards

The environmental benefits and hazards of tissue culture technology are not expected to be different from those due to the development of new varieties of crops by conventional breeding alone. Utilization of tissue culture would merely shorten the time needed for development.

Development of new crops from weed species is desirable for numerous reasons (Princen, 1977):

> To reduce dependency on foreign imports for a variety of raw materials such as natural rubber and vegetable oils
>
> To reduce shortages of specialty chemicals, such as essential oils, tannins, and alkaloids, and other materials, such as waxes, fibers, fats and oils
>
> To provide new protein sources for livestock feed in areas where established crops cannot grow
>
> To increase land utilization by developing species adapted to marginal conditions
>
> To develop new speciality crops for small farm regions throughout the world that are not amenable to monoculture

Increased competition with conventional crops is likely to be the major environmental concern for most weed or nuisance species considered for crop development. As a result, more herbicides would have to be used, and new selective herbicides would have to be developed. In previously undeveloped areas, competition between superior varieties of weed crops and the native flora could produce changes in species composition and succession, and ultimately, a reduction in the stability of the ecosystem. This may be particularly true in fragile ecosystems, such as desert, alpine, and coastal regions. However, the likelihood that there will be increased competition seems very slim especially since modifications of the weed species for use in conventional agricultural systems will probably render them less competitive.

6

Stakeholders

DOMESTIC ACTIVITIES

Immense excitement has been generated in recent years by the advent of recombinant DNA technology and the prospect that applied genetics will improve the quality of life in many ways. This interest arose from findings made in basic research labs at universities, which quickly burgeoned into a multimillion dollar commercial industry. Activities on both fronts are expanding continuously. Meanwhile, various government agencies have developed an interest in this area owing, in part, to concerns for public safety arising from overly fast commercialization of a technology whose safety has not been established absolutely. Thus, all three sectors–universities, private industry, and government–are deeply interested in the evolution of the applied genetics field. From a socioeconomic viewpoint, applied genetics will provide the opportunity to analyze and improve relationships between industries and universities on the one hand, and between industries and government on the other.

Universities

Most fundamental advances in both the science and engineering aspects of biotechnology have been made in university research labs. This fact will continue to hold true for some time to come, although considerable expertise is now being acquired by commercial firms engaged in applied genetics R&D.

A few of the many academic scientists who have contributed to the foundation of the applied genetics industry are listed in Table 6.1. This list, by no means exhaustive, includes many of those prominent academic

Table 6.1: A Few Academic Scientists Engaged in Genetic Engineering Research; Commercial Affiliations (not inclusive)

Name	University	Affiliation
Bert O'Malley	Baylor	
James Bonner	Cal Tech	
Leroy Hood	Cal Tech	
Gerald Fink	Cornell	Advisor to Collaborative Genetics
Walter Gilbert	Harvard	Co-founder of Biogen
Philip Leder	Harvard	
Tom Maniatis	Harvard	
Matthew Meselson	Harvard	
Mark Ptashne	Harvard	Founder of Genetics Institute
Dan Nathans	Johns Hopkins	Advisor to Monsanto
Hamilton Smith	Johns Hopkins	Advisor to Cetus
David Baltimore	MIT	Advisor to Collaborative Genetics
David Botstein	MIT	Advisor to Collaborative Genetics
Arnold Demain	MIT	Advisor to Cetus
Philip Sharp	MIT	Co-founder of Biogen
Paul Berg	Stanford	
Stanley Cohen	Stanford	Advisor to Cetus
Ronald Davis	Stanford	Advisor to Collaborative Genetics
Roy Curtiss	Univ. of Alabama	
Martin Cline	UCLA	Co-founder of AMgen
Winston Salser	UCLA	Co-founder of AMgen
John Baxter	UCSF	
Herbert Boyer	UCSF	Co-founder of Genentech
Anand Chakrabarty	Univ. of Illinois	Advisor to Petrogen
David Jackson	Univ. of Michigan	On leave to Genex
Stanley Falkow	Univ. of Washington	Advisor to Cetus
Winston Brill	Univ. of Wisconsin	Advisor to Cetus
Timothy Hall	Univ. of Wisconsin	Advisor to Agrigenetics
Howard Temin	Univ. of Wisconsin	
Frank Ruddle	Yale	

scientists who have become affiliated with one or another genetic engineering firm. Several companies were founded through the efforts and energies of university researchers who, nevertheless, maintained faculty status at their academic institutions. This state of affairs has occasioned a certain degree of rivalry among university scientists who now view their research as potentially lucrative. As a result, the qualities of cooperation and intercommunication that once characterized academic research have been seriously compromised. This trend is likely to continue for the foreseeable future with accompanying improvements in industry-university relations at the expense of freedom of information flow within the scientific community. The situation could improve if private industries undertake programs to support basic academic research on an unrestricted no-strings-attached basis. Commercial firms are being encouraged by Congress to do so via proposed tax credits and other investment incentives. Corporate backing of academic research has recently become especially desirable since federal sources of funds for basic biomedical research (i.e., NIH and NSF) have failed to keep pace with growing demand.

University faculties are organizing to offer their services as technical experts in applied genetics. Two examples are:

> Bioinformation Associates, Inc.–A group of MIT professors from the biology, chemistry, and chemical engineering departments who provide wide-ranging consulting services for basic and applied research in genetic engineering.

> Biotechnology Research Center–Established at Lehigh University in Bethlehem, Pennsylvania, this joint effort of scientists and engineers provides education in biotechnology and conducts research in the areas of biomass conversions, microbial desulfurization of coal, and improved methods for waste treatment.

Harvard University recently developed, then rejected, a plan to establish its own genetic engineering company. This proposal evolved as a means to put the considerable talent of the Harvard faculty to the purpose of generating profits for the university, rather than to serve the interests of outside commercial firms (such as Biogen, co-founded by Harvard biologist, Walter Gilbert). The plan succeeded only in generating controversy. Faculty members argued that profit motives would add to the rivalry that already existed within the biology department and that traditional academic goals of education and research are incompatible with a profit-making orientation. Eventually, the plan was abandoned, but Harvard biologist Mark Ptashne, who conceived the venture, proceeded to establish his own firm, called Genetics Institute, Inc., located in nearby Somerville, Massachusetts. Harvard considered, then declined, an offer to acquire 10% equity in this new firm.

In California, engineering departments of Stanford University and the University of California at Berkeley have joined forces to establish a nonprofit biotechnology center that has acquired a 30% stake in Engenics Inc., a newly incorporated genetic engineering company specializing in continuous fermentation processes and equipment. The company has already received a $7½ million infusion of capital from six major corporations: Bendix, Elf-Aquitaine, General Foods, Koppers,

Maclaren Power & Paper, and Mead. The two universities will channel any capital appreciation or stock dividends arising from their shareholdings in Engenics into support for research in the chemical engineering and medical microbiology departments at each campus.

Commercial Firms

The excitement generated by the field of biotechnology, particularly recombinant DNA and genetic engineering, has been felt most emphatically in the private sector of the U.S. economy. We have identified over 200 companies currently engaged in some aspect of modern applied genetics. More firms are becoming involved every month. It is estimated that capital investment in applied genetics R&D reached $500 million in 1980. In five more years, the value will be $5 billion, and in ten years, $25 billion. Many investors and business analysts anticipate that the decade of the 1980s will occasion a "biology boom" akin to the electronics explosion of the 1970s.

Table 6.2: Specialist Companies*

Name Location	Business	Corporate Partners or Investors
Abec Fermentations Allentown, PA	Process equipment	
Advanced Genetic Sciences Greenwich, CT	Plant genetics	Rohm & Haas Cardo (Sweden)
Advanced Genetics Research Institute Berkeley, CA	Plant genetics	
Advanced Mineral Technology Socorro, NM	Mineral leaching	
Agri-Business Research Scottsdale, AZ	Plant genetics Desert plants	Arizona St. Univ.
Agrigenetics Denver, CO	Plant genetics	
AgroBiotics Baltimore, MD	Plant genetics Biomedical	
Almar Boston, MA	Animal vaccines	
Alpha Therapeutic Los Angeles, CA	Biomedical Vaccines	Green Cross (Japan)
American Diagnostics Newport Beach, CA	Hybridomas	

(continued)

Table 6.2: (continued)

Name Location	Business	Corporate Partners or Investors
Angenics Cambridge, MA	Hybridomas	
Applied Biosystems Foster City, CA	Instrumentation	
Applied Genetics Boston, MA	Biomedical	
Applied Molecular Genetics (AMGen)	Biomedical	Abbott Labs TOSCO
Armos San Francisco, CA	Instrumentation	
Associated Biomedic Systems Buffalo, NY	Biomedical	
Atlantic Antibodies Bar Harbor, ME	Hybridomas	
ATP-Syntro San Diego, CA	Biofuels	
Bethesda Research Labs Gaithersburg, MD	Biomedical	
Bioassay Systems Woburn, MA	Diagnostics	
BioCell Technology New York, NY	Diagnostics	
Biochemical Corp. of America Birmingham, AL	Pollution control	Division of Sybron
BioChem Technology Malvern, PA	Instrumentation	
Bio-Con Bakersfield, CA		
Bio-Gas Arvada, CO	Biofuels	
Biogenex Laboratories Dublin, CA	Biomedical Hybridomas	
BioInformation Assoc. Boston, MA	Consulting	
Biolaffite Princeton, NJ	Instrumentation	Biolaffite (France)

(continued)

Table 6.2: (continued)

Name Location	Business	Corporate Partners or Investors
Bio-Response Wilton, CT	Biomedical Cell culture	
Biosearch San Rafael, CA	Instrumentation Biomedical	
Biotec Madison, WI	Enzymes	
Biotech Framingham, MA	Biomedical	Dennison, Framingham, MA
Biotech Research Labs Bethesda, MD	Biomedical Cell culture	Ethyl, Richmond, VA
Biotechnica Cambridge, MA		
Biotechnica International Princeton, NJ		
Bio-Technical Resources Manitowoc, WI	Consulting	
Bioz Arlington, VA	Enzymes Pesticides	
Brain Research New York, NY	Diagnostics	
Braun San Francisco, CA	Instrumentation	
Calgene Davis, CA	Plant genetics	Allied Corp.
Cellbiology Cold Spring Harbor, NY	Biomedical	
Centaur Genetics Chicago, IL	Hybridomas	
Centocor Philadelphia, PA	Hybridomas	FMC, Philadelphia
Cetus Berkeley, CA	Biomedical Energy Agriculture	National Distillers Chevron, Amoco Shell Oil
Chemapec Woodbury, NY	Instrumentation	
Chiron Emeryville, CA	Biomedical Vaccines, hormones	Elf-Aquitaine (France)
Clonal Research Newport Beach, CA	Hybridomas	

(continued)

Table 6.2: (continued)

Name Location	Business	Corporate Partners or Investors
Codon Brisbane, CA		
Collaborative Genetics Waltham, MA	Biomedical	Dow Chemical Green Cross (Japan)
Collagen Corp. Palo Alto, CA	Biomedical	Monsanto
Creative Agri-Management Champaign, IL	Biofuels Ethanol from biomass	
Cytogen Princeton, NJ	Hybridomas	
Cytox New York, NY	Pollution control	
Damon-Biotech Needham Heights, MA	Drug delivery	Damon Corp.
DNA Plant Technology Cinnaminson, NJ	Plant genetics	Campbell Soup
DNAX Institute Palo Alto, CA	Biomedical	Division of Schering- Plough
Ecoenergetics Fairfield, CA	Chemicals Energy	
Electro-Nucleonics Fairfield, NJ	Diagnostics Vaccines	
EMV Rockville, MD	Computer applica- tions	
Energetics Palo Alto, CA	Instrumentation	
Engenics Menlo Park, CA	Process engineering Instrumentation	Bendix, Koppers Maclaren Power & Paper Mead, Elf-Aquitaine General Foods
Envirotech Salt Lake City, UT	Mineral leaching	
Enzo Biochem New York, NY	Biomedical	Enzo Japan
Enzyme Center Boston, MA	Enzymes	
Fermentech Los Gatos, CA	Instrumentation	
Flow General McLean, VA	Biomedical Cell culture	

(continued)

Table 6.2: (continued)

Name Location	Business	Corporate Partners or Investors
Geneco San Francisco, CA	Instrumentation	
Genentech San Francisco, CA	Biomedical	Fluor, Eli Lilly Monsanto, Toray (Japan) Lubrizol, Hoffmann-LaRoche, IMC
General Environmental Science Beachwood, OH	Pollution control	
Genetic Design Watertown, MA	Instrumentation	
Genetic Diagnostics Great Neck, NY	Hybridomas	
Genetic Engineering Denver, CO	Animal genetics Embryo transplants	Immuno Genetics, Boston
Genetic Oil Co. (Genoco) New York, NY	Biofuels	
Genetic Replication Technologies Newport Beach, CA		
Genetic Systems Seattle, WA	Hybridomas	Syntex
Genetics Institute Boston, MA	Biomedical	
Genetics International Boston, MA		
Genex Rockville, MD	Biomedical Amino acids	Monsanto, Koppers Bristol-Myers
Geno Southfield, MI	Biomedical	
Genzyme Boston, MA	Enzymes	
Griffen Sciences Hartford, CT	Insecticides	
HEM Research Rockville, MD	Biomedical Cell culture	
Hybridoma Sciences Atlanta, GA	Hybridomas	Emery University
Hybrigen Los Angeles, CA	Hydridomas	

(continued)

Table 6.2: (continued)

Name Location	Business	Corporate Partners or Investors
Hybritech La Jolla, CA	Hybridomas	Mitsubishi (Japan)
Immulok Carpinteria, CA	Hybridomas	
Immunex Seattle, WA	Hybridomas	
ImmunoGenetics Boston, MA	Hybridomas	Division of Genetic Engineering, Denver
Immunorex Philadelphia, PA	Hybridomas	FMC/Centocor Venture
Immunotec Tampa, FL	Hybridomas	
Immunotech Cambridge, MA	Hybridomas	
Immutron Newport Beach, CA	Hybridomas	Nuclear Medical Systems
Imreg New Orleans, LA	Diagnostics	Safeguard Scientifics, King of Prussia, PA
Integrated Genetics Framingham, MA	Biomedical	
IntelliGenetics	Instrumentation	Schlumberger
Interferon Sciences New York, NY	Biomedical	National Patent Development Corp., NY
International Genetic Sciences Jamaica, NY		
International Genetics (INGENE) Santa Monica, CA	Plant genetics	
International Plant Research Institute (IPRI) San Carlos, CA	Plant genetics	Eli Lilly
Kallestad Laboratories Austin, TX	Instrumentation	
Kappa Scientific Escondido, CA	Instrumentation	
Key Interferon Tampa, FL	Biomedical	Southern Medical and Pharmaceutical, FL
Lee Biomolecular San Diego, CA	Biomedical	

(continued)

Table 6.2: (continued)

Name Location	Business	Corporate Partners or Investors
Life Sciences St. Petersburg, FL	Biomedical	
Liposome Labs Princeton, NJ	Drug delivery	
Liposome Technology Menlo Park, CA	Drug delivery	Cooper Labs
M.A. BioProducts Walkersville, MD	Diagnostics	Division of Whittaker
Meloy Laboratories Springfield, VA	Plant genetics	Division of Revlon
Microlife Technics Sarasota, FL	Biomedical Cell culture	
Molecular Diagnostics New Haven, CT	Hybridomas	
Molecular Genetics Minnetonka, MN	Biomedical Animal vaccines	American Cyanamid
Monoclonal Antibodies Palo Alto, CA	Hybridomas	
Native Plants Salt Lake City, UT	Plant genetics	
NeoBionics Boston, MA	Hybridomas	
New England BioLabs Beverly, MA	Enzymes	
New England Nuclear Boston, MA	Hybridomas	Division of DuPont
Ocean Genetics Walnut Creek, CA		
Oncogene Seattle, WA	Biomedical	
Petrogene New York, NY	Biofuels	
Petroleum Fermentations (PETROFERM) New York, NY	Biofuels	
Phyllogenics San Francisco, CA	Plant genetics	
Phytogen Pasadena, CA	Plant genetics	

(continued)

Table 6.2: (continued)

Name Location	Business	Corporate Partners or Investors
Phytotech Houston, TX	Plant genetics	
Phyto-Tech Labs Torrance, CA	Plant genetics	
Plant Genetics Davis, CA	Plant genetics	INCO (Canada)
Plant Tissue Culture San Mateo, CA	Plant genetics Exotic plants	
Polybac Allentown, PA	Pollution control	Division of Cytox
Quadroma Escondido, CA	Hybridomas	Johnson & Johnson
Quantum Biotech Labs Alhambra, CA		
R&A Plant/Soils Pasco, WA	Soil conditioning	
Repligen Boston, MA	Biomedical	Gillette
Replitech Petaluma, CA		
Ribi Immunochem Research Hamilton, MT	Biomedical	
Rio Vista Genetics San Antonio, TX	Animal genetics Embryo transplants	

Salk Institute
  Biotechnology/Industrial
  Assoc. (SIBIA)
La Jolla, CA
	Biomedical	Phillips Petroleum

Salsbury Labs Charles, LA		
Southern Biotech Tampa, FL	Biomedical	
Sungene San Francisco, CA	Plant genetics	Lubrizol
Swine Genetics San Francisco, CA	Animal genetics	
Synergen Ft. Collins, CO	Biomedical Chemicals	

(continued)

Table 6.2: (continued)

Name Location	Business	Corporate Partners or Investors
Syngene West Lafayette, IN		
Syngenics Santa Clara, CA		
Sys-Tec New Brighton, MA	Instrumentation	
Tech America Elwood, KA	Animal genetics	
Transformation Research Framingham, MA		
Unigene Laboratories Totowa, NJ	Biomedical	
University Genetics Norwalk, CT	Funds university R&D Patent licensing	Division of University Patents
Vega Biotechnologies Tucson, AZ	Instrumentation	
Viragen Miami, FL	Biomedical	
Viralab Emeryville, CA	Biomedical	
Viratek Covina, CA	Biomedical	Division of ICN
Virogenetics New York, NY	Biomedical	
Wescor Logan, UT	Instrumentation	
Xenogen Mansfield, CT		
Zarcon Berkeley, CA		
Zoecon Palo Alto, CA	Insecticides	Division of Occidental Petroleum
Zymark Hopkinton, MA	Instrumentation	
Zymos Seattle, WA		

*Based in part on information supplied by S. King, F. Eberstadt Inc.

In general, capital investment in biotechnology has occurred along two different paths. Initially small, new companies specializing in genetic engineering (see Table 6.2) were created by young scientists/businessmen who combined keen foresight with a propensity for financial risk-taking. These individuals anticipated the huge commercial potential of modern biological techniques and managed to attract venture capital to underwrite their business plans. The two preeminent examples of this venture capital approach are:

> Cetus Corp., founded in 1971 by UC Berkeley physicist Donald A. Glaser, biochemist Ronald E. Cape (who also earned an MBA degree from Harvard), and Peter J. Farley (a medical doctor with an MBA from Stanford). Even before the advent of recombinant DNA techniques, Cetus funded its operations through contracts with larger commercial firms, especially pharmaceutical houses. Current backers of Cetus include major oil companies, such as Amoco, Chevron, and Shell.

> Genentech, Inc., founded in 1976 by Robert A. Swanson (who holds degrees in chemistry and business management from MIT) and UCSF biochemist Herbert W. Boyer. The firm was established expressly to commercialize on DNA technology and was initially underwritten by venture capital, chiefly from Kleiner & Perkins in California, Wilmington Securities in Delaware, and Lubrizol Enterprises in Ohio. Genentech currently operates with capital derived from specific contracts with large firms, such as Hoffmann-LaRoche and Eli Lilly, and with capital derived from a recent public sale of stock.

The contribution of small, innovative companies such as these to the emergence of the applied genetics industry has been summarized by Nelson M. Schneider, a drug-industry investment analyst for E.F. Hutton:

> All major new technologies have been promoted and fostered by small companies. The small guys have the opportunity only because the bigger guys ignore it. The big companies can't see the forest for the trees. They choose not to participate because of their own ingrown bureaucracies.

Thus, small companies such as Cetus and Genentech, and the many dozens of similar young firms that have sprung up recently, have the flexibility and the expertise to take advantage of scientific advances in the applied genetics field. But today's small companies are determined to grow. Says Peter Farley, president of Cetus:

> It's biology's turn now. We actually saw it coming, and we were determined right from the outset to become a major company. It's not a get-rich-quick scheme. We expect to be around fifty years from now as a major company.

It remains to be determined whether Cetus, as a major corporation, can continue to "see the forest for the trees."

The second principal business strategy for investing in biotechnology has been for large, technically oriented companies to undertake independent, in-house R&D programs (see Table 6.3). Virtually every major U.S. pharmaceutical firm has engaged in or made plans to initiate recombinant DNA research. The high level of interest among firms in this industry stems from the obvious applications of the new technology to the manufacturing of new or improved drugs. Some drug firms have undertaken collaborative research ventures with small genetic engineering firms directed towards the development of specific products. Examples include arrangements between Genentech and Hoffmann-LaRoche to make interferon, Genentech and Eli Lilly to make human insulin, and Genex and Bristol-Myers to make interferon.

Table 6.3: U.S. Industrial Companies

Name Location	Projects	Partners
I. Agriculture, food and forestry industries		
Archer-Daniels-Midland Decatur, IL	Biofuels Ethanol from corn	
Campbell Soup Camden, NJ	Vegetables	DNA Plant Technology
CPC International Englewood, NJ	Biofuels Ethanol from corn	
Crown Zellerbach San Francisco, CA	Forestry	
DeKalb AgResearch DeKalb, IL	Crop seeds	
Frito-Lay Dallas, TX	Potatoes, corn	
General Mills Minneapolis, MN	Wheat Food processing	
Pioneer Hi-Bred Des Moines, IA	Crop seeds	

(continued)

Table 6.3: (continued)

Name Location	Projects	Partners
Ralston Purina St. Louis, MO	Food processing	
A.E. Staley Decatur, IL	Corn products Fructose syrup	
Weyerhauser Centralia, WA	Forestry	

II. Petrochemical industry

Name Location	Projects	Partners
Allied Morristown, NJ	Industrial and agri-chemicals	BioLogicals (Canada) Calgene
American Cyanamid Wayne, NJ	Biomedical Industrial chemicals	Molecular Genetics
ARCO Los Angeles, CA	Oil applications Pollution control	
Ashland Oil Ashland, KY	Biofuels Ethanol from bio-mass	Publicker Industries
Celanese New York, NY	Biofuels Enzymes	Yale University
Diamond Shamrock Dallas, TX	Oil applications Desert plants	Univ. of Arizona
Dow Chemical Midland, MI	Biomedical Industrial chemicals	Collaborative Genetics
Ethyl Richmond, VA	Biomedical	Biotech Research Labs
Exxon New York, NY	Oil applications	
FMC Philadelphia, PA	Biomedical Industrial chemicals	Centocor
Goodyear Tire & Rubber Akron, OH	Desert plants (guayule)	
W.R. Grace New York, NY	Industrial chemicals	
Koppers Pittsburgh, PA	Industrial chemicals Process engineering Mining applications	Engenics Genex
Lubrizol Wickliffe, OH	Oil applications	Genentech

(continued)

Table 6.3: (continued)

Name Location	Projects	Partners
Monsanto St. Louis, MO	Biomedical Plant genetics Industrial and agri- chemicals	Biogen (Switzerland) Collagen Corp. Genentech Genex
National Distillers and Chemicals New York, NY	Process engineering	Cetus
Occidental Petroleum Los Angeles, CA	Oil applications Insecticides	Zoecon
Pennzoil Houston, TX	Oil applications Desert plants	
Phillips Petroleum Bartlesville, OK	Biomedical Oil applications	Salk Institute
P.L. Biochemicals Milwaukee, WI	Biomedical Amino acids	
Rohm & Haas Philadelphia, PA	Agrichemicals	Advanced Genetic Sciences
Shell Oil Houston, TX	Biomedical Oil applications Industrial and agri- chemicals	Cetus
Standard Oil (Calif.) San Francisco, CA	Oil Applications Industrial chemicals Fructose syrup	Cetus
Standard Oil (Ind.) Chicago, IL	Oil Applications Xanthan gums	Cetus
Standard Oil (Ohio) Cleveland, OH	Oil applications	
Stauffer Chemical Westport, CT	Industrial chemicals	
TOSCO Los Angeles, CA	Oil applications	AMGen
Union Carbide Danbury, CT	Industrial chemicals	
UOP Des Plaines, IL	Industrial chemicals Microbial catalysts	National Oil Co. of Spain
III. Pharmaceutical industry		
Abbott Labs North Chicago, IL	Biomedical	AMGen

(continued)

Table 6.3: (continued)

Name Location	Projects	Partners
ACS Pharmaceuticals Edina, MN	Biomedical Hybridomas	
Baxter-Travenol Labs Deerfield, IL	Diagnostics Enzymes	
Becton-Dickinson Rutherford, NJ	Diagnostics Instrumentation	
Bristol-Myers New York, NY	Biomedical	Genex
Cooper Labs Palo Alto, CA	Diagnostics Drug delivery	Liposome Technology
ICN Pharmaceuticals Covina, CA	Biomedical	Viratek
Johnson & Johnson New Brunswick, NJ	Biomedical	BioLogicals (Canada) Quadroma
Eli Lilly Indianapolis, IN	Biomedical	Genentech IPRI
Litton Bionetics Rockville, MD	Biomedical Agrichemicals	
Mallinckrodt St. Louis, MO	Hybridomas	Washington Univ.
McNeil Pharmaceutical Spring House, PA	Diagnostics	
Merck Rahway, NJ	Vaccines Diagnostics	
Miles Labs Elkhart, IN	Enzymes Citric acid	
Ortho Pharmaceutical Raritan, NJ	Diagnostics	Division of Johnson & Johnson
Pennwalt Philadelphia, PA	Drug delivery	
Pfizer New York, NY	Biomedical Agrichemicals	
Revlon New York, NY	Cosmetic applications	Meloy Labs
Richardson-Merrell Wilton, CT	Biomedical	Division of Dow Chemical
Schering-Plough Kenilworth, NJ	Biomedical	Biogen (Switzerland)
G.D. Searle Skokie, IL	Biomedical	

(continued)

Table 6.3: (continued)

Name Location	Projects	Partners
Smith Kline/Beckman Philadelphia, PA	Biomedical Instrumentation	
Squibb Princeton, NJ	Biomedical	
Upjohn Kalamazoo, MI	Biomedical	
Warner-Lambert Morris Plains, NJ	Diagnostics	

IV. Pollution control, process engineering industries

Name Location	Projects	Partners
Battelle Columbus Columbus, OH	Pollution control Microbial detoxica- tion	
Bio-Rad Labs Richmond, CA	Instrumentation	
Corning Glass Corning, NY	Biomedical Instrumentation	
Davy-Mckee Chicago, IL	Process engineering	Cetus
Dynatech R&D Cambridge, MA	Biofuels Instrumentation	
Eastman-Kodak Rochester, NY	Diagnostics Instrumentation	
Fluor Irvine, CA	Process engineering	Genentech
General Electric Schenectady, NY	Pollution control	
Jacobs Engineering Pasadena, CA	Pollution control	
Arthur D. Little Cambridge, MA	Biofuels	U.S. Dept. of Energy
SRI International Stanford, CA	Pollution control Microbial detoxica- tion	

Large corporations representing other industrial sectors are also investing heavily in applied genetics. DuPont, the world's largest chemical producer, has undertaken a sizeable commitment to R&D in the biosciences. Likewise, several major oil companies, such as Amoco and Phillips, have established in-house programs in biotechnology. These large, wealthy companies are hiring high-quality scientists and bioengineers for the purpose of developing biological solutions for problems such as alternative sources of energy and petrochemical feedstocks. Collaborative agreements with small firms exist here too, such as contracts between Cetus and several oil companies, including Amoco, Chevron, and Shell, to conduct R&D on energy-related projects.

While small genetic engineering firms will continue to conduct laboratory-scale R&D, commercial scale-up of biotechnological processes will require capital investment that only large firms can undertake. Thus, the relative importance of large companies, with respect to the growth of the applied genetics industry, will increase at the expense of the smaller firms. A trend can be anticipated paralleling that which occurred in the semiconductor industry during the 1970s; namely, larger firms will acquire through merger (or drive out of business) the many small, specialized genetic engineering companies that have emerged.

The recent efforts of Cetus and Genentech to raise large sums of money by offering shares of stock to the public reveal the difficulty that small companies face when undertaking capital-intensive projects. A short time ago, the corporate management of both companies expressed desires to avoid going public with their stock until the mid-1980s at the earliest. However, several factors served to alter their plans: (1) commitments to pursue costly in-house programs of commercial scale-up; (2) the failure to attract additional financing from large corporate backers (such as Chevron and Amoco in the case of Cetus); and (3) the absence to date of saleable products derived from R&D investments. Public ownership may compel these companies to lose some of their flexibility and farsightedness that provides them with the competitive edge over large, bureaucratic corporations.

Federal Government Activities

The involvement of the federal government in applied genetics stems from a concern, first expressed by research scientists in the mid-1970s, that the application of recombinant DNA techniques could produce new organisms that might escape from the laboratory and endanger the human population and the environment. Thus, the role of the government in biotechnology has so far been limited to considering the practice of recombinant DNA methods in academic and commercial settings.

The National Institutes of Health: In the United States, the National Institutes of Health (NIH) control the administration of all federally supported recombinant DNA research and of all such activities carried out by commercial firms in voluntary compliance with NIH Guidelines for Research Involving Recombinant DNA Molecules. Other government agencies also involved in the potential regulation or control of commercial activities are the Food and Drug Administration (FDA), the Occupational Safety and Health Administration (OSHA), the National Institute for Occupational Safety and Health (NIOSH), and the Environmental Protection Agency (EPA). Figures 6.1 and 6.2 depict the organizational relationships between the various government agencies and any company involved in recombinant DNA activities.

The purpose of the NIH guidelines for recombinant DNA research is to specify proper practices for constructing and handling recombinant DNA molecules and for handling organisms and viruses containing such molecules. Compliance with the guidelines is mandatory for all institutions engaging in such research and receiving federal support. The guidelines were first published in the *Federal Register* in the summer of 1976. Since then they have been amended considerably and now reflect a more confident and relaxed attitude about potential risks inherent in these activities.

The director of the NIH is responsible for the establishment, implementation, and final interpretation of the guidelines. Pursuant to the guidelines, the Director has established the Recombinant DNA Advisory Committee (RAC) and the Office of Recombinant DNA Activities (ORDA) to provide technical and administrative assistance in the fulfillment of these responsibilities (Figure 6.2).

The RAC was established to provide technical and scientific assistance to the Director of NIH. Consequently, its membership must collectively reflect expertise in scientific fields relevant to recombinant DNA technology and biological safety. Additonally, at least 20% of its members must be knowledgeable about applicable law, standards of professional conduct and practice, the environment, public and occupational health, and related fields. The RAC meets four times a year and advises the NIH Director on changing the containment levels specified for various types of experiments covered under the guidelines, assigning containment levels to experiments not covered by the guidelines, and recommending new host-vector systems. The recombinant DNA field has expanded rapidly over the past few years, especially with the increasing involvement of private industry. Consequently, the RAC has been compelled to assess large-scale fermentation procedures, to examine confidential industrial data, and to review occupational safety and health standards.

However, the RAC has little expertise in industrial engineering. Furthermore, as an advisory committee to a nonregulatory agency, it has no authority to require compliance with its advice. The RAC has recently decided to limit its assessment of industrial facilities to an examination only of the biological characteristics of the operation (45 *FR* 77379). Consequently, private commercial firms will no longer be requested to submit to the NIH the details of their physical plants, medi-

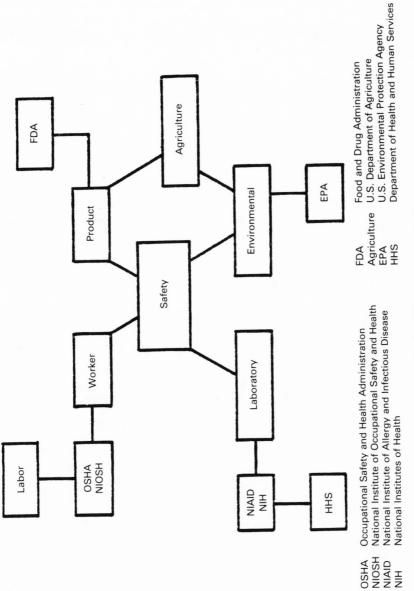

Figure 6.1: Federal interagency coordination.

OSHA   Occupational Safety and Health Administration
NIOSH   National Institute of Occupational Safety and Health
NIAID   National Institute of Allergy and Infectious Disease
NIH   National Institutes of Health

FDA   Food and Drug Administration
Agriculture U.S. Department of Agriculture
EPA   U.S. Environmental Protection Agency
HHS   Department of Health and Human Services

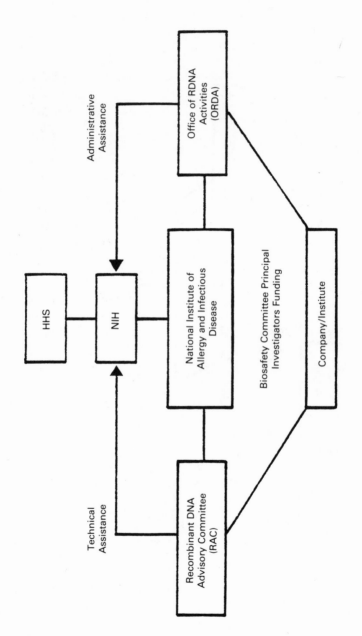

Figure 6.2: Laboratory safety.

cal surveillance programs, environmental monitoring schemes, or emergency plans.

The Office of Recombinant DNA Activities (ORDA) is responsible for providing maximum access to information on every aspect of the recombinant DNA field. As the focal point for all information on such activities, it provides technical and administrative advice to any institution, agency, or individual within or outside the NIH. In addition, it serves as executive secretary to the RAC, publishes the *Recombinant DNA Technical Bulletin,* and reviews and approves IBC membership lists (see below). ORDA's responsibilities also include the scheduling and announcing of RAC meetings and the publishing of revised guidelines in the *Federal Register.* ORDA also distributes information regarding policy decision relevant to recombinant DNA research, announcements of training courses dealing with experimental and safety issues, up-dating of approved host-vector systems and other experimental protocols, and publishes a bibliography of newly released articles on recombinant DNA.

INFORMATION SOURCES

A number of newsletters have been started which provide analyses and digests of scientific articles, patent filings, financial occurrences and general news. As can be seen in Table 6.4, most of the newsletters are highly specialized.

Table 6.4: Genetic Engineering Newsletters and Directories

NEWSLETTERS

Name	Publisher	Price
Agricultural Genetics Report	Mary Ann Liebert New York, NY (bimonthly)	$90/yr
Applied Genetics News	Business Communications Co., Stamford, CT (biweekly)	$200/yr
Bioengineering News	Thomas Mysiewicz San Francisco, CA (biweekly)	$295/yr
Biotechnology Bulletin	Scientific & Technical Studies London, UK (monthly)	$160/yr

(continued)

NEWSLETTERS (continued)

Biotechnology Law Report	Mary Ann Liebert New York, NY (in preparation)	?
Biotechnology News	CTB International Summit, NJ (biweekly)	$185/yr
Biotechnology Newswatch	McGraw-Hill New York, NY (biweekly)	$377/yr
Biotechnology Press Digest	Mary Ann Liebert New York, NY (monthly)	$185/yr
Genetic Engineering Letter	Gershon Fishbein Washington, DC (biweekly)	$295/yr
Genetic Engineering News	Mary Ann Liebert New York, NY (monthly)	$90/yr
Genetic Technology News	Technical Insights Inc. Fort Lee, NJ (monthly)	$132/yr

DIRECTORIES

Biotechnology USA	IMSworld Publications London, UK	$500
Biotechnology Europe	IMSworld Publications London, UK	$400
Biotechnology Japan	IMSworld Publications London, UK	$300
Genetic Engineering Biotechnology Firms: USA, 1981	Noyes Data Corp.	$100
Genetic Engineering Industry Director	International Resource Development Inc. Norwalk, CT	$640
TELEGE: Biotechnology and Bioscience Director	Environment and Information Center, Inc. New York, NY	?

The NIH guidelines were amended in January, 1980 to include a section dealing with voluntary compliance by private commercial firms engaging in recombinant DNA activities. This action was taken as a compromise to proposed mandatory controls put forward by the FDA. A scheme of voluntary compliance encourages private companies to follow the same administrative and technical procedures that are required of any federally supported institution. However, all items of information provided to the RAC or to ORDA by complying companies are protected as trade secrets, thus prohibiting subsequent disclosure under the Freedom of Information Act. In addition, companies that comply voluntarily are asked to register all projects involving recombinant DNA technology.

In April 1980, the NIH published a set of guidelines dealing with large-scale applications of recombinant DNA methodology. These guidelines detail the physical containment requirements for the production of recombinant DNA organisms in volumes exceeding ten liters.

Any company engaged in recombinant DNA activities and in voluntary compliance with the NIH guidelines must establish an Institutional Biosafety Committee (IBC). Each firm's IBC must include at least five members, with a minimum of two (but no fewer than 20%) having no affiliation with the company. The IBC members collectively must have expertise in recombinant DNA research and be capable of assessing the safety of such experiments and the risks to workers and the community. The nonaffiliated members are to represent the best interests of the surrounding community with respect to health and the protection of the environment. Officials of state or local public health or environmental protection agencies, members of other local government organizations or persons active in community medical, occupational health, or environmental affairs are all eligible to serve on an IBC.

The primary responsibility of the IBC is to review all recombinant DNA experiments conducted by the company to ensure compliance with the NIH guidelines. The IBC review must include an assessment of the containment levels utilized as well as an evaluation of the facilities, procedures, training, and expertise of the personnel conducting the experiments. The IBC must also adopt emergency plans covering accidental spills and contamination resulting from recombinant DNA research.

In addition to an IBC, any firm or institution engaging in recombinant DNA R&D involving high levels of physical containment (P3 or P4) must appoint a biological Safety Officer (BSO). The BSO is a member of the IBC and is responsible for conducting periodic inspections of lab facilities, reporting to the IBC any significant violations of the guidelines, developing emergency plans, and interacting with the principal investigator (PI) in areas of lab security, technical and safety procedures, and adherence to the guidelines.

There has been considerable debate within the RAC, and within the scientific community at large, whether the NIH guidelines serve a useful purpose. Support is growing for total elimination of even the minimal restrictions now imposed by the guidelines. Others argue that the absence of federal policy resulting from such action would lead to a profusion of fragmentary regulations at the local level that may prove more inhibitory than the current federal guidelines. It is clear that a consensus has emerged that recombinant DNA experimentation imposes risks to the public that are no greater than those arising from general microbiological research and that, if federal guidelines and local IBCs are to be maintained, these "pseudo-regulatory" authorities had better examine a broad range of health and safety issues related to biological R&D and refrain from singling out recombinant DNA activities for such surveillance.

Other Federal Agencies: Recombinant DNA technology is rapidly moving out of the exclusive domian of university research laboratories and into industrial laboratories and large-scale production facilities. Concurrently, government agencies other than the NIH are becoming increasingly involved with issues concerning environmental monitoring, worker safety and health, and product quality.

The Food and Drug Administration has been involved from the very beginning with industrial scale-up of recombinant DNA technology. This interest stems from the fact that the first commercial products emerging from this new technology are likely to be intended for human use; namely, insulin, human growth hormone, and interferon.

New drugs intended for human use must be certified by the FDA through the approval of two company-submitted forms: (1) a notice of Claimed Investigation Exemption for a New Drug (IND); and (2) a New Drug Application (NDA). Together these forms supply the FDA with proprietary information on drug composition, results of human and animal testing, and manufacturing procedures. As of June 1980, the position of the FDA was that drugs produced by recombinant DNA technology could not be marketed under existing INDs or NDAs as simply changes in manufacturing technique.

Submission of an IND informs the FDA that a company has tested a potential new drug and that it will be testing it further. Required by the form is a statement of the methods, facilities, and controls used for the manufacturing, processing, and packaging of the new drug to establish and maintain appropriate standards of identity, strength, quality, and purity.

The NDA is a request for approval to market the drug. Although intended primarily to provide information on the results of clinical testing, the NDA also contains detailed information on the manufacturing of the drug. It covers all the information that the sponsor knows about the drug and often consists of thousands of pages.

The Occupational Safety and Health Administration (OSHA) was established under the Occupational Safety and Health Act of 1970. OSHA is a regulatory agency within the Department of Labor that is charged with developing and promulgating standards, formulating and enforcing appropriate regulations to maintain safe and healthful conditions in the workplace. OSHA has announced that it will develop a recombinant DNA regulatory policy over the next two years. This could prove to be a difficult task since hazards associated with this technology have remained speculative. Two recent events will further contribute to the difficulty that OSHA will encounter in efforts to regulate recombinant DNA technology. In a recent Supreme Court decision, reduced standards for exposure to benzene were disallowed owing to a lack of evidence that the existing exposure levels were dangerously high. Similarly, there exists no firm evidence of risk resulting from contact with recombinant DNA organisms (at any level of exposure). Secondly, OSHA has no authority to preview the technical details that a company intends to use in the large-scale production of recombinant DNA organisms. Until now, OSHA has obtained this information from the RAC, but, as mentioned above, the RAC no longer intends to gather this information.

The National Institute for Occupational Safety and Health (NIOSH) was established by the Occupational Safety and Health Act of 1970. NIOSH is a component of the Center for Disease Control under the Public Health Service and is authorized to conduct research and recommend workplace standards to OSHA. NIOSH is interested in the following areas relevant to recombinant DNA:

> Process operations with attendant potential for worker exposure

> Engineering controls, such as physical containment design, ventilation, exhaust gas filtration, waste product control, etc.

> Validation procedures pertaining to sterilization of equipment, physical containment, and process termination

> Work practices, emergency and accident procedures, medical surveillance, environmental monitoring, and employee training and education

Patent Issues: The issue of patent protection for products and processes evolving from recombinant DNA R&D is both controversial and very important to commercial firms engaged in these activities. Two events have had a sizeable impact so far. One established the legal precedent that man-made microorganisms are not excluded from patent protection; the second extended patent coverage to the inventors of certain basic laboratory procedures.

On June 16, 1980, the U.S. Supreme Court ruled in a narrow 5-4 decision that the General Electric Corp. should not be denied patent protection on an "oil-eating" microorganism developed by Dr. A.M. Chakrabarty. The decision hinged on whether a microorganism is unpatentable subject matter simply because it is alive. The Court found that the principal criteria upon which an invention is deemed patentable (namely, that it be new, useful, and nonobvious) were in no way infringed by the fact that the invention is alive. The dissenting minority argued that those who originally framed the patent statutes never intended that patent protection be afforded to living things. The Court admitted to a lack of competence in evaluating the potential dangers or benefits of this new technology, and they further declared that any binding policy regarding patentability of living organisms must originate in Congress.

On December 2, 1980, Stanford University and the University of California were jointly awarded a patent dealing with gene cloning techniques used in recombinant DNA experiments. The techniques, developed by Stanley Cohen at Stanford and Herb Boyer at UCSF, have become the basis for virtually all recombinant DNA experimentation to date. The two universities have declared that they will license the technology to any company that wishes to employ the techniques and they will collect royalties on its use. They have further stated that a condition for use of the technology will be adherence to the NIH guidelines. However, the patent applies only to the use of recombinant DNA technology within the borders of the United States and would not cover overseas operations by U.S.-based companies, or by foreign firms. It is highly unlikely that this patent could withstand a legal challenge if it is deemed to inhibit the commercial development of recombinant DNA technology. Moreover, litigants will argue that numerous refinements to the basic techniques have been made so that the original inventors are no longer entitled to patent protection.

FOREIGN ACTIVITIES

The overseas practice of applied genetics has proceeded in a fashion similar to its evolution in the United States. Much of the basic biological research that gave rise to this new industry occurred in foreign laboratories, particularly in Western Europe. As in the United States, there has emerged in several countries a variety of small new genetic engineering companies. Likewise, established corporations are engaging in applied genetics R&D (see Table 6.5).

Table 6.5: Foreign Companies*

Name	Projects	Partners
Canada		
Allelix	Diverse	Labatt Brewery
BioLogicals	Biomedical Instrumentation	Allied (U.S.) Johnson & Johnson (U.S.)
Connaught Labs	Biomedical	
INCO	Biomedical Mining applications	Biogen (Switz.)
Levochem Industries	Amino acids	Division of Commercial Organics (Canada)
Philom Bios	Animal Health Fermentations	
Sybron Biochemical	Pollution control	Paper Research Inst. of Canada
United Kingdom		
ABM Chemicals	Enzymes	
Alcon Biotechnology	Diverse	
Biozyme	Enzymes	
Burroughs-Wellcome	Biomedical Tissue culture	
Cambridge Life Sciences	Enzymes	
CellTech	Diverse	British & Commonwealth Shipping 100% owned by British government
Floranova	Plant genetics	
Fospur	Pollution control	
HTL Specialties	Industrial chemicals	Tate & Lyle (U.K.) Hercules (U.K.)
Imperial Chemical Industries (ICI)	Single-cell protein	
Monotech	Hybridomas	
Sera-Lab	Hybridomas	
France		
Chimie Industrielle	Industrial chemicals	
Elf-Aquitaine	Oil Applications Biofuels Amino acids Process engineering	Chiron (U.S.) Engenics (U.S.) Transgene (France)

(continued)

Table 6.5: (continued)

Name	Projects	Partners
G3 (Groupement Genie Genetique)	Biomedical	Institut Pasteur INSERM CNRS INRA
Genetica	Chemicals	Rhone-Poulenc (France)
Immunoteque	Hybridomas	INSERM
LaFarge-Coppee	Amino acids Fermentations	
Rhone-Poulenc	Biomedical Industrial chemicals	Genetica (France)
Sanofi	Biomedical	Institut Pasteur
Transgene	Biomedical	Elf-Aquitaine (France) Moet-Hennessey (France) BSN (France)
Switzerland		
Bioengineering	Fermentations	
Biogen	Biomedical	Grand Metropolitan (U.K.) INCO (Canada) IMC (U.S.) Monsanto (U.S.) Schering-Plough (U.S.)
Ciba-Geigy	Biomedical Agrichemicals Plant genetics	
Chemap	Diverse	Cardo (Sweden)
Hoffmann-LaRoche	Biomedical	Genentech (U.S.)
Sandoz	Biomedical Agriculture and foods applications	
Other European		
X (Belgium)	Human genetic screening	
Novo Industri (Denmark)	Biomedical Enzymes	
Medix (Finland)	Hybridomas	
Gist-Brocades (Holland)	Biomedical Enzymes	

(continued)

Table 6.5: (continued)

Name	Projects	Partners
Nizo (Holland)	Biofuels	Dutch Institute for Dairy Research
Alfa-Laval (Sweden)	Biofuels	
Hilleshog (Sweden)	Agriculture applications	Cardo (Sweden)
Kabi Vitrum (Sweden)	Biomedical	Genentech (U.S).
Boehringer-Mannheim) (West Germany)	Biomedical Enzymes	
Hoechst (West Germany)	Biomedical	Massachusetts General Hospital (U.S.)
Schering (West Germany)	Biomedical	
Israel		
Biotechnology General	Agriculture applications	Elron (Israel)
Interpharm Labs	Biomedical	Ares R&D (Switz.) Yeda R&D (Israel)
Koor Food	Single-cell protein	Yeda R&D (Israel)
Japan		
Ajinomoto Kanegafuchi Chemical Kikkoman Shoyu Kyowa Hakko Kogyo Mitsubishi Chemical Mitsubishi Petrochemical Mitsui Toatsu Chemicals Nippon Kayaku Nippon Shokubai Kagaku Shionogi Showa Denko Sumitomo Chemical Suntory Takeda Chemical Toray Industries	Recombinant DNA applications	
Asahi Chemical Dainippon Ink & Chemical Green Cross Kirin Brewery Mochida Pharmaceutical Oji Paper Toyo Jozo	Cell culture technologies	

(continued)

Table 6.5: (continued)

Name	Projects	Partners
Daicel Chemical		
Denki Kagaku Kogyo		
Hiki		
Kao Soap		
Mitsubishi Gas & Chemical	Bioreactor develop-	
Mitsubishi Kakoki Kaisha	ment	
Nippon Oil		
Sanyo Chemical		
Tanabe Seiyaku		
Unitika		
Australia		
Australian Monoclonal	Hybridomas	
Bioclone Australia	Hybridomas	
Biotechnology Australia	Diverse	CRA Ltd. (Australia)

*Based in part on information supplied by S. King, F. Eberstadt Inc.

In contrast to U.S. activities, however, some foreign governments have supplied considerable financial backing to fledgling genetic engineering companies. For example, the British government, in concert with four London investment firms, has established Celltech. This nationally owned venture came into being only after extensive hand-wringing on the part of government planners, but Celltech can now hope to commercialize significant scientific achievements of British researchers, several of whom have already lost the opportunity to capitalize on their findings owing to a lack of public interest. (For example, the monoclonal antibody technique was discovered in England, but the scientists involved failed to patent the process within the necessary time limits.)

With regard to applied genetics in France, the government there has fostered a healthy relationship between universities and industries, thereby facilitating the transfer of basic biotechnology from academic labs into the commercial sector. The French government is committed to spend about $25 million over the next five years in support of biotechnology. Several French government science and research agencies have cooperated in support of a new business venture, called G3, which will concentrate on biomedical applications of genetic engineering.

In Japan, where government-industry cooperation is legendary, over a hundred established chemical and pharmaceutical firms are actively pursuing genetic engineering programs with government support. The Japanese are considered world leaders in certain areas of biotechnology, particularly fermentation techniques. They are far ahead of the rest of the world with regard to the quantity and diversity of products, such as antibiotics, vitamins, and food additives, that can be readily manufactured by fermentation procedures.

Most of the commercial development of genetic engineering in Canada has proceeded via private investment. A small new firm, BioLogicals in Toronto, recently signed a multimillion dollar agreement with Allied Chemical (a U.S. firm) to conduct applied research into the uses of genetic engineering for the production of industrial and agricultural chemicals. Connaught Laboratories, formerly associated with the University of Toronto and the site where the hormone insulin was first isolated in 1921, has been largely taken over by the Canadian government. The firm is now engaged in an ambitious revitalization program that includes large-scale investment in genetic engineering.

In Israel, biotechnology is being applied to meet national needs in the areas of agriculture, industrial chemicals, and waste management. Considerable effort is being expended to investigate various types of photosynthetic algae as potential sources of single-cell protein and useful biochemicals. Genetically engineering salt tolerance into algae, thus allowing the microbes to thrive in brackish ponds, has received special attention.

For the most part, government regulation (or pseudo-regulation) of recombinant DNA activities in foreign countries has followed a path similar to that in the United States. Actual legislation dealing with this area exists only in the United Kingdom, where the GMAG (Genetic Manipulation Advisory Group) reviews experimental protocols much as does the RAC in this country. In Britian, however, emphasis has been placed solely on physical containment of recombinant DNA organisms, rather than on both physical and biological containment, as in the United States. Other Western European nations have generally followed the model set by GMAG.

The Japanese government has followed the U.S. lead in establishing voluntary guidelines for recombinant DNA research. The trend in Japan, as in all nations, has been to continually revise downward the restrictions imposed by the guidelines as information accumulates indicating biohazards inherent to recombinant DNA techniques are no greater than the risks associated with microbiological methods in general. Governments in all nations are fearful that unnecessary regulation of genetic engineering may adversely affect the commercial potential that this new technology offers.

Glossary

This list of definitions is intended to help the reader and should not be considered all inclusive.

Aerobe
Organism requiring oxygen.

Anaerobe
Organism able to live in the absence of oxygen; some anaerobes are "obligate;" i.e., they are killed in the presence of oxygen.

Antigen
Any chemical substance, natural or man-made, that elicits an immune response in animals.

Bacteriophage
One of a subgroup of viruses that infect bacteria; consists of a relatively small amount of DNA contained in a protein coat.

Callus
Tissue growing from the disorganized proliferation of cells excised from a plant.

Chloroplast
A cellular organelle in higher plants; site of photosynthesis.

Chromosome
The basic macrostructure of heredity; organization of DNA in cell nuclei containing large numbers of genes.

164

Clone
A collection of cells each having an identical genetic composition.

Codon
A triplet of nucleotides on a DNA chain that specifies a particular amino acid or otherwise controls protein synthesis.

Colicin
A bacterial toxin, the coding for which is found on a plasmid; some forms are toxic to humans.

Conjugation
One-way transfer of DNA between bacteria in cell contact.

Crown gall
Plant tumor caused by infection with *Agrobacterium tumefaciens;* genes located in the Ti-plasmid of the *Agrobacterium* are responsible for tumor induction.

DNA
Deoxyribonucleic acid; the molecular basis of genes; made from the sequential arrangement of four nucleotide building blocks: adenine, cytosine, guanine, and thymine; normal configuration is in double-stranded helical form.

cDNA
Complementary DNA; laboratory-created DNA that is complementary to mRNA extracted from a cell.

Entomopathogen
Insect pathogen, usually microbial in nature, such as a bacterium, protozoan, or virus.

Enzyme
Organic catalyst of biochemical reactions in a cell; composed of protein.

Eukaryote
An organism composed of cells that are distinguished by the presence of a nucleus and multiple chromosomes; fungi, protozoa, and all differentiated multicellular forms of life are eukaryotic.

F factor
Fertility factor; plasmid that specifies gender in bacteria.

Gene

A defined length along a chromosome, made of DNA and coding for a protein molecule.

Gene library

The result of a shotgun experiment in which each cloned bacterial colony contains different segments of DNA.

Genome

All the genes of an organism or individual.

HEPA filter

High efficiency particulate air filter.

Host-vector system

In the recombinant DNA field, the particular organism (host) into which the gene is cloned, and the vehicle (vector)–usually a plasmid system–that carries the gene into the host.

Intron

An intervening sequence of DNA of unknown function found only in eukaryotic genes; this sequence is not expressed in the transcription to mRNA.

Lac operon

An operon in *E. coli* that codes for three enzymes involved in the metabolism of lactose.

Lambda

Bacteriophage that infects *E. coli*; commonly used as a vector in recombinant DNA research.

Ligase

An enzyme that catalyzes the linking of sequential bases in single-stranded DNA.

Lignocellulose

Complex biopolymer comprising the bulk of woody plants; consists of polysaccharides and polymeric phenols.

Lymphocyte

A type of cell found in the blood, spleen, lymph nodes, etc., of higher animals; one sub-class of lymphocyte manufactures and secretes antibodies.

Lysis
  Process of cell disintegration; cell bursting.

Nonconjugable
  Refers to bacterial plasmids that cannot be transferred between organisms.

Nucleotide
  Any of a class of compounds consisting of a purine or pyrimidine base, bonded to a ribose or deoxyribose sugar and to a phosphate group; the basic structural units of RNA and DNA.

Nucleus
  The cell region containing chromosomes and enclosed in a definite membrane; found only in eukaryotic cells.

Oligonucleotide
  The sequential arrangement of more than one nucleotide.

Operator
  A region of DNA that controls the expression of adjacent genes by interacting with a repressor protein.

Operon
  A gene unit consisting of one or more genes and the controls for that unit; the lac operon, for example, is made up of three genes, the operator, and the terminator.

Opines
  Unusual amino acids synthesized by genes located on Ti-plasmids; nopaline and octopine are examples.

Peptide
  Two or more amino acids joined together.

Phage
  Shortened form of the word bacteriophage; bacterial virus.

Plasmid
  A small circle of double-stranded DNA that exists and replicates autonomously in bacteria; often codes for resistance to antibiotics; may be transferred between bacteria during conjugation.

Polymerase

Enzyme that catalyzes the assembly of nucleotides into RNA and of deoxynucleotides into DNA.

Polypeptide

A molecular chain of many amino acids joined together; synonymous with protein.

Prokaryote

Cellular organism distinguished by the lack of a defined nucleus and by the presence of a single, naked chromosome; bacteria and blue-green algae are the only major examples.

Promoter

The region on a DNA strand that indicates the place to start the transcription of the gene into RNA.

Protein

A sequence of amino acids; the ultimate expression of a gene; primary component of enzymes and many hormones.

Protoplast

A bacterium or plant cell from which the cell wall has been removed.

Regeneration

Manipulation of individual cells or masses of cells to grow into whole plants.

Replication

The process of making copies of DNA in a cell; depending upon the plasmid, many copies may be replicated after insertion of DNA into the host.

Repressor

A gene product that prevents the transcription of an operon by binding to an operator region.

Restriction endonuclease

One of a class of enzymes that cleave both strands of DNA at sequence-specific sites; used extensively in recombinant DNA experiments.

R factor

Resistance factor; refers to plasmids coding for resistance to antibiotics.

Ribosome
A large molecular array, composed of RNA and protein, that is responsible for translating messenger RNA into protein.

RNA
Ribonucleic acid, used to form complements to DNA in gene expression.

mRNA
Messenger RNA, formed in the cell nucleus in the process of gene expression; complementary to the base sequence of the DNA of the gene.

Shotgun cloning
Cloning procedure in which all the chromosomes of a donor organism are enzymatically fragmented and placed into hosts for expression; results in a gene library.

Sticky end
Refers to double-stranded DNA that has been cleaved in such a way by certain restriction endonuclease enzymes that the end of one strand extends beyond the end of the other; the end is called "sticky" because the bases are exposed and can thus mate with complementary sticky ends.

T-DNA
A region of the Ti-plasmid that contains genes required for crown gall tumor induction and maintenance.

Terminator
A region on a gene that codes for the termination of transcription.

Ti-plasmid
A large plasmid found in *Agrobacterium tumefaciens;* induces crown gall tumors in plants infected with the bacterium.

Transcription
The process of copying DNA into RNA; the result is messenger RNA.

Transduction
The transfer of genetic material from one cell to another by means of a viral vector (for bacteria, the vector is bacteriophage).

Transformation

The process of inserting into the host organism a vector containing a gene that is to be cloned.

Translation

The process of making a peptide from mRNA; performed by ribosomes.

Transposons

Short DNA segments containing one or a few genes that are readily translocated between cells or to different sites within the same cell; responsible for antibiotic resistance in bacteria; also found in some eukaryotic organisms; transposons may serve as suitable vectors for genetic engineering in various organisms.

Vector

An agent consisting of a DNA molecule known to autonomously replicate in a cell to which another DNA segment may be attached experimentally to bring about the replication of the attached segment.

Virus

Any of the submicroscopic infective agents composed of RNA or DNA wrapped in a protein coat and capable of growth and multiplication only in living cells.

Bibliography

I. GENERAL

Abbott, A.J., 1978, "Practice and promise of woody species," *Acta Horticult.,* 79:113-127.

Allen, G., and Fantes, K.H., 1980, "A family of structural genes for human lymphoblastoid (leukocyte-type) interferon," *Nature,* 287:408-11.

Bahl, C.P., Marians, K.J., and Wu, R., 1976, "A general method for inserting specific DNA sequences into cloning vehicles," *Gene,* 1:81-92.

Barker, S.A., and Somers, P.J., 1978, "Biotechnology of immobilized multienzyme system," *Adv. Biochem. Eng.,* 10:27-49.

Barz, W., Reinhard, E., and Zenk, M.H. (eds.), 1977, *Plant Tissue Culture and Its Biotechnology Application,* Springer-Verlag, NY.

Beers, R.F., and Bassett, E.G., 1977, *Recombinant Molecules: Impact on Science and Society,* Raven Press, NY.

Binding, H., 1980, "Isolated plant protoplasts in genetics and plant breeding," *Theor. Appl. Genet.,* 56:90.

Bull, A.T., Ellwood, D.C., and Ratlege, C. (eds.), 1979, *Microbial Technology: Current State, Future Prospects,* Cambridge Univ. Press, NY.

Cape, R.E., 1979, "The industrial revolution in microbiology," *Med. Progr.,* 34:1619-1623.

Chakrabarty, A.M., ed., 1978, *Genetic Engineering,* CRC Press, Boca Raton, FL.

Cline, M.J., Stang, H., Mircola, K., Morse, L., Ruprecht, R., Browne, J., and Salser, W., 1980, "Gene transfer in intact animals," *Nature,* 284, 422-426.

Cohen, S.N., 1975, "The manipulation of genes," *Scient. Amer.,* 233:25-33.

Cohen, S.N., Chang, A.C.Y., Boyer, H.W., and Helling, R.B., 1973, "Construction of biologically functional bacterial plasmids *in vitro,*" *Proc. Nat. Acad. Sci.,* U.S.A., 70:3240-3244.

Colijn, C.M., Kool, A.J., and Nijkamp, H.J.J., 1979, "An effective chemical mutagenesis procedure for *Petunia* hybrid cell suspension cultures," *Theor. Appl. Genet.,* 55:101-106.

Cozzarelli, N.R., Melechen, N.E., Jovin, T.M., and Kornberg, A., 1967, "Poly-nucleotide cellulose as a substrate for a polynucleotide ligase induced by phage T4," *Biochem. Biophys. Res. Commun.*, 28:578-586.

Crea, R., Kraszewski, A., Hirose, T., and Itakura, K., 1978, "Chemical synthe-sis of genes for human insulin," *Proc. Nat. Acad. Sci., USA*, 75:5765-5769.

Gefter, M.L., Becker, A., and Hurwitz, J., 1967, "The enzymatic repair of DNA," *Proc. Nat. Acad. Sci., USA*, 58:240-247.

Gilbert, W., and Villa-Komaroff, L., 1980, "Useful proteins from recombinant bacteria," *Scient. Amer.* 242:74-94.

Grobstein, C., 1979, "The recombinant DNA debate," *Scient. Amer.*, 237:22-36.

Hishinuma, F., Tanaka, T., and Sakaguchi, K., 1978, "Isolation of extrachromo-somal DNA from extremely thermophilic bacteria," *J. Gen. Microbiol.*, 104:193-199.

Itakura, K., and Riggs, A.D., 1980, "Chemical DNA synthesis and recombinant DNA studies," *Science*, 209:1401-1405.

Jackson, D.A., Symons, R.H., and Berg, P., 1972, "Biochemical method for in-serting new genetic information into DNA of simian virus 40," *Proc. Nat. Acad. Sci., USA*, 69:2904-2909.

Kennedy, J.F., 1979, "Facile methods for the immobilization of microbial cells without disruption of their life processes," *Am. Chem. Soc. Symp. Series*, 106:119-132.

Kennett, R.H., McKearn, T.J., and Bectheol, K.B. (eds.), 1980, *Monoclonal Antibodies*, Plenum Press, NY.

Marmur, J., 1961, "A procedure for the isolation of DNA from microorganisms," *J. Molec. Biol.*, 3:208-218.

Milstein, C., 1980, "Monoclonal antibodies," *Scient. Amer.* 243:66-74.

Morgan, J., and Whelan, W.J., (eds.), 1979, *Recombinant DNA and Genetic Ex-perimentation*, Pergamon Press, NY.

Mulligan, R.C., and Berg, P., 1980, "Expression of a bacterial gene in mam-malian cells," *Science*, 209:1422-1427.

Novick, R.P., 1980, "Plasmids," *Scient. Amer.*, 243:102-127.

Office of Technological Assessment, 1980, *Impacts of Applied Genetics*, Wash-ington, D.C.

Peppler, H.J., and Perlman, D. (eds.), 1979, *Microbial Technology*, 2nd Ed., Academic Press, NY.

Perlman, D., 1974, "Prospects for the fermentation industries, 1974-1983," *Chemtech*, 4:210-216.

Richards, J. (ed.), 1978, *Recombinant DNA: Science, Ethics, and Politics*, Aca-demic Press, NY.

Reinert, J., and Bajaj, Y.P.S., 1977, *Applied and fundamental aspects of plant cell, tissue, and organ culture*, Springer-Verlag, NY.

Schaffner, W., 1980, "Direct transfer of cloned genes from bacteria to mam-malian cells," *Proc. Nat. Acad. Sci., USA*, 77:2163-2167.

Sebek, D.K., and Laskin, A.I. (eds.), 1979, *Genetics of Industrial Microorgan-isms*, Am. Soc. Microbiol., Washington, D.C.

Skinner, K.J., 1975, "Enzymes technology," *Chem. Eng. News*, 53:22-41.

Svoboda, A. 1978, "Fusion of yeast protoplast induced by polyethylene glycol," *J. Gen. Microbiol.*, 109:169-175.

Vasil, I.K., Ahuja, M.R., and Vasil, V., 1979, "Plant tissue cultures in genetics and plant breeding," *Adv. Genet.*, 20:127-215.

Wade, N., 1980, "UCLA gene therapy racked by friendly fire," *Science*, 210:509-511.

Wetzel, R., 1980, "Applications of recombinant DNA technology," *Amer. Scient.*, 68:664-675.

Wingard, L.B., Katchalski-Katzir, E., and Goldstein, L. (eds.), 1979, *Enzyme technology*, Academic Press, NY.

II. PHARMACEUTICAL

Bell, G.I., Swain, W.F., Pictect, R., Cordell, B., Goodman, H.M., and Rutter, W.J., 1979, "Nucleotide sequence of a cDNA clone encoding human prepro-insulin," *Nature*, 282:525-527.

Bibb, M., Schottel, J.L., and Cohen, S.N., 1980, "A DNA cloning system for interspecies gene transfer in antibiotic-producing *Streptomyces*," *Nature*, 284:526-531.

Bloom, B.R., 1980, "Interferons and the immune system," *Nature*, 284:593-595.

Cape, R.E., 1979, "Microbial genetics and the pharmaceutical industry," *Chemtech*, 9:638-644.

Chemical Week, February 6, 1980, "Gene feat spurs interferon race."

Derynck, R., Remant, E., Saman, E., Stansses, P., De Clercq, E., Content, J., and Fiers, W., 1980, "Expression of human fibroblast interferon gene in *E. coli*," *Nature*, 287:193-197.

Fiddeo, J.C., Seeburg, P.H., Denoto, F.M., Hallewell, R.A., Baxter, J.D., and Goodman, H.M., 1979, "Structure of genes for human growth hormone and chorionic somatomammotropin," *Proc. Nat. Acad., USA*, 76:4294-4298.

Forbes Magazine, January 5, 1981, "Drugs."

Goeddel, D.V., et al., 1980, "Human leukocyte interferon produced by *E. coli* is biologically active," *Nature*, 287:411-416.

Goeddel, D.V., Heyneker, H.L., Hozumi, T., Arentzen, R., Itakura, K., Yansura, D.G., Ross, M.J., Miozzari, G., Crea, R., and Seeburg, P.H., 1979, "Direct expression in *E. coli* of a DNA sequence coding for human growth hormone," *Nature*, 281:544-548.

Goeddel, D.V., Kleid, D.G., Bolivar, F., Heyneker, H.L., Yansura, D.G., Crea, R., Hirose, T., Kraszewski, A., Itakura, K., and Riggs, A.D., 1979, "Expression in *E. coli* of chemically synthesized genes for human insulins," *Proc. Nat. Acad. Sci., USA*, 76:106-110.

Henriquez, P., Candia, A., Norambuena, R., Silva, M., and Zemelman, R., 1979, "Antibiotic properties of marine algae," *Bot. Mar.*, 22, 451-454.

Kieslich, K., 1980, "New examples of microbial transformations in pharmaceutical chemistry," *Bull. Soc. Chim. Fr.*, 112:9-17.

Lorz, H., and Potrykus, I., 1979, "Regeneration of plants from mesophyll protoplasts of *Atropa belladonna*," *Experientia*, 35:313-314.

Marx, J.L., 1980, "Interferon congress highlights," *Science*, 210:998.

Miller, H.I., Guerigirian, J.L., Troendle, G., and Sobel, S., 1980, "Aspects of the drug regulatory process: recombinant DNA technology," *Recomb. DNA Techn. Bull.*, 3:72-74.

Miozzari, G., 1980, "Strategies for obtaining expression of peptide hormones in *E. coli," Recomb. DNA Techn. Bull.,* 3:57-67.

Ross, M.J., 1980, "Production of medically important polypeptides using recombinant DNA technology," *Recomb. DNA Techn. Bull.,* 3:1-11.

Shine, J., Fettes, I., Lan, N.C.Y., Roberts, J.L., and Baxter, J.D., 1980, "Expression of cloned beta-endorphin gene sequences by *E. coli," Nature,* 285: 456-461.

Shiner, G., 1980, "Human growth hormone: potential for treatment are broadened," *Res. Resour. Report.,* 4:1-5.

Sun, M., 1980, "Insulin wars: new advances may throw market into turbulence," *Science,* 210:1225-1228.

U.S. Environmental Protection Agency, 1976, *Pharmaceutical Industry Hazardous Waste Generation, Treatment, and Disposal,* SW-508, Washington, D.C.

Valenzuela, P., Gray, P., Quiroga, M., Zaldivar, J., Goodman, H.M., and Rutter, W.J., 1979, "Nucleotide sequence of the gene coding for the major protein of hepatitis B virus surface antigen," *Nature,* 280:815-819.

Woodruff, H.B., 1980, "Natural products from microorganisms," *Science,* 208: 1225-1229.

III. CHEMICAL

Buchanan, R.A., Cull, I.M., Otey, F.H., and Russell, C.R., 1978, "Hydrocarbon and rubber producing crops: evaluation of U.S. species," *Econ. Botany,* 32:131-145.

Buchta, K., 1974, "Biotechnical production of organic acids," *Chem. Zeit.,* 98: 532-538.

Chemical Week, June 4, 1980, "Enzymes are a sweet way to do business."

Chemical Week, October 8, 1980, "Biotechnology: research that could remake industries."

Chemical Week, March 4, 1981, "What applied genetics might do in chemicals."

Johnson, J.D., and Hinman, C.W., 1980, "Oil and rubber from arid land plants," *Science,* 208:460.

Khafagy, S.M., Metwally, A.M., Eil-Ghazooly, M.G., and El-Naggar, S.F., 1979, "Sesquiterpene lactones from *Varthemia candicans," Planta Med.,* 37:75-78.

Markwell, A.J., 1978, "Some chemical processes involving microorganisms," *Chemsa,* 4:44-45.

Miwa, T.K., 1979, "Chemicals bloom in the desert," *Chemical Week,* 124:31-33.

Nyiri, L., 1971, "Preparation of enzymes by fermentation," Intern. *Chem. Eng.,* II:447-458.

Pape, M., 1976, "The competition between microbial and chemical processes for the manufacture of basic chemicals and intermediates," Sem. on Microb. Energy Conversion, United Nations Inst. for Training and Research, October, 1976.

Sanderson, J.E., Wise, D.L., and Augenstein, P.C., 1979, "Organic chemicals and liquid fuels from algal biomass," *Biotechnol. Bioeng. Symp.,* 8:131-151.

Schwartz, R.D., Williams, A.L., and Hutchinson, D.B., 1980, "Microbial production of 4,4-dihydroxy biphenyl: hydroxylation by fungi," *Appl. Environ. Microbiol.,* 39:702-708.

Tilak, B.D., 1978, "Prospect of manufacture of industrial chemicals from cellulosic raw materials," *Symp. Proc. Bioconversion Cellulosic Substances, New Delhi, February, 1977.*

Wang, D.I.C., Cooney, C.L., Demain, A.L., Gomez, R.F., and Sinskey, A.J., 1978, "Degradation of cellulosic biomass and its subsequent utilization for production of chemical feedstocks," MIT Program Rep. No. COO/4198-6.

Yoshiharu, I., Ichino, C., and Tamis, I., 1978, "Production and utilization of amino acids," *Angew. Chem.,* 17:176-183.

IV. ENERGY

Benemann, J.R., and Hallenbeck, P.C., 1978, "Recent developments in hydrogen production by microalgae," Symp. on Energy from Biomass and Wastes, Inst. Gas Technol., Chicago, IL.

Chin, K.K., and Gohr, T.N., 1978, "Bioconversion of solar energy: methane production through water hyacinth," from Symp. on Energy from Biomass and Wastes, Inst. of Gas Technol., Washington, D.C., p. 215.

Clausen, E.C., Sitton, O.C., and Gaddy, J.L., 1979, "Biological production of methane from energy crops," *Biotechnol. Bioeng.,* 21:1209-1219.

Da Silva, E.J., 1980, "Biogas: fuel of the future?" *Ambio,* 9:2.

Dunlop, D.D., 1976, "Microbial oil recovery," from Sem. on Microb. Energy Conversion, United Nations Inst. for Training and Research, October, 1976.

Gerson, D.F. and Zajic, J.E., 1979, "Bitumen extraction from Athabasca tar sands with microbial surfactants," *Petroleum Abstract,* 19(32), No. 266,277.

Gulf Oil Chemicals Co., 1979, "Biomass feedstocks of the future," *Processing,* 25:38-39.

Hall, D.O., Reeves, S.G., Dennis, G., and Rao, K.K., 1978, "Biocatalytic hydrogen production," from *Conf. on Sun: Mankind's Future Source of Energy, New Delhi,* Vol. 2, p. 805.

Hashimoto, A.G., Chen, Y.R., and Prior, R.L., 1979, "Methane and protein production from animal feedlot wastes," *J. Soil and Water Conservation,* 34:16.

Keenan, J.D., 1979, "Review of biomass to fuels," *Proc. Biochem.,* 14:9-12.

Khan, A.W., 1979, "Anaerobic degradation of cellulose by mixed culture," *Can. J. Microbiol,* 23:1700-1705.

King, S.R., 1979, "Gasohol: ethanol from plant matter as motor fuel," F. Eberstadt & Co., Inc., NY.

Lasking, A.I., 1979, "Microbial transformations of hydrocarbons," *174th Am. Chem. Soc. Mtg.,* 24:848-850.

Loehr, R.C., 1978, "Methane from human, animal, and agricultural wastes" in: *Renewable Energy Resources and Rural Applications in the Developing World,* Westview Press, Boulder, CO.

Lonsane, B.K., Singh, H.D., and Baruah, J.N., 1976, "Use of microorganism and microbial products in secondary recovery of petroleum from economically unrecoverable oil reservoirs," *J. Scient. Industr. Res.,* 35:316.

Morris, W., and Whiteley, M., 1978, "Liquid fuels from carbonates by a microbial system," *Am. Chem. Soc. Symp. Series,* 90:120-132.

Pankhurst, E.S., 1980, "Biogas," Gas Eng. Mang., 20:3.

Pimentel, L., and Calvin, M., 1979, "Brazil's biomass program is one of the most extensive," *Chem. Eng. News,* 57:35.

Reed, T.B., 1975, "Biomass energy refiners for production of fuel and fertilizers," *J. Appl. Polymer Sci.,* 28:1-9.

Schwab, C., 1979, "Energy from vegetation: legal issues in biomass energy conversion," *Solar Law Reporter,* 1:784.

Seeley, J.Q., 1974, "Geomicrobiological method of prospecting for petroleum," *Oil Gas J.,* 72:142-144.

Sitton, O.C., and Gaddy, J.L., 1979, "Design and performance of an immobilized cell reactor for ethanol production," from 72nd Ann. Mtg. Am. Inst. Chem. Eng., San Francisco, CA, Abstract No. 41.

Smith, G.D., 1978, "Microbiological hydrogen production," *Search,* 78:209.

Tornabene, T.G., 1977, "Microbial formation of hydrocarbons," in *Proc. Symp. on Microbial Energy Conversions, Goettingen, Germany,* Pergamon Press, NY.

U.S. Environmental Protection Agency, 1979, "Process design manual; sludge treatment and disposal," EPA 625/1-29-001, September, 1979.

Yen, T.F., 1976, "Microbial oil shale extraction," from Seminar on Microbial Energy Conversion, United Nations Inst. for Training and Research, October, 1976.

Yen, T.F., and Meyer, W.C., 1976, "Enhanced dissolution of oil shale by bioleaching with *Thiobacilli*," *Appl. Environ. Microbiol.,* 32:610-616.

Zajic, J.E., Kosaric, N., and Brosseau, J.D., 1978, "Microbial production of hydrogen," *Adv. Biochem. Eng.,* 9:57-109.

V. MINING

Bruynesteyn, A., and Duncan, D.W., 1971, "Microbiological leaching of sulphide concentrates," *Canad. Metal Quart.,* 10:57-63.

Duncan, D.W., and Bruynesteyn, A., 1971, "Enhancing bacterial activity in a uranium mine," *Canad. Min. Metal. Bull.,* 74:116-120.

Duncan, D.W., Landesman, J., and Walden, C.C., 1967, "Role of *Thiobacillus ferrooxidans* in the oxidation of sulfide minerals," *Can. J. Microbiol.* 13:397-403.

Duncan, D.W., and Walden, C.C., 1972, "Microbiological leaching in the presence of ferric iron," *Develop. Indust. Microbiol.,* 13:66-75.

Gates, J.E., and Pham, K.D., 1979, "An indirect fluorescent antibody staining technique for determining population levels of *Thiobacillus ferrooxidans* in acid mine drainage waters," *Microb. Econ.,* 8:121-128.

McGoran, C.J.M., Duncan, D.W., and Walden, C.C., 1969, "Growth of *Thiobacillus ferrooxidans* on various substrates," *Can. J. Microbiol.,* 15:135-138.

Murr, L.E., Torma, A.E., and Brierly, J.A. (eds.), 1978, *Metallurgical Applications of Bacterial Leaching and Related Microbiological Phenomena,* Academic Press, NY.

Razzell, W.E., and Trussel, P.C., 1963, "Isolation and properties of an iron-oxidizing *Thiobacillus*," *J. Bacteriol.,* 85:595-603.

Sakaguchi, H., and Silver, M., 1976, "Microbiological leaching of a chalcopyrite concentrate by *Thiobacillus ferrooxidans*," *Biotechnol. Bioeng.*, 18:1091-1101.
Torma, A.E., Walden, C.C., and Branion, R.M.R., 1970, "Microbiological leaching of a zinc sulfide concentrate," *Biotechnol. Bioeng.*, 12:501-517.

VI. POLLUTION CONTROL

Alexander, M., 1981, "Biodegradation of chemicals of environmental concern," *Science*, 211:132-138.
Bellinick, C., Batistic, L., and Mayadon, J., 1979, "Degradation of 2,4-D in the soil," *Rev. Ecol. Biol. Sol.* (France), 16:161-168.
Brown, M.J. and Lesfer, J.N., 1979, "Metal removal in activated sludge: the role of bacterial extracellular polymers," *Water Res.*, 13:817-838.
Chakrabarty, A.M., Friello, D.A., and Bopp, L.N., 1978, "Transposition of the plasmid DNA segments specifying hydrocarbon degradation and their expression in various microorganisms," *Proc. Nat. Acad. Sci., USA,* 15:3109-3112.
Chemical Week, July 23, 1980, "Building 'superbugs' for the big cleanup."
Crawford, R.L., 1977, "Novel methods for enumeration and identification of microorganisms for potential use in biological delignification," from Symp. on Biological Delignification, Weyerhaeuser, August, 1976, pp. 55-72.
Davis, A.J., and Yen, T.F., 1976, "Feasibility studies of a biochemical desulfurization method," *Am. Chem. Soc. Symp.* 74:137.
Deschamps, A.M., Mahoudeau, G., Conti, M., and Lebeault, J.M., 1980, "Bacteria degrading tannic acid and related compounds," *J. Ferment. Technol.,* 5:93-97.
Detz, C.M., and Barvinchak, G., 1979, "Microbial desulfurization of coal," *Mineral Cong. J.,* 65:75-82.
Finnerty, W.R., 1980, "Microbial desulfurization and denitrogenation," 180th Am. Chem. Soc. Mtg., Las Vegas, NV.
Grady, C.P.L., and Grady, J.K., 1979, "Industrial wastes: fermentation industry," *J. Water Pollut. Contr. Fed.,* 81:1325.
Harbold, H.S., 1976, "How to control biological waste treatment processes," *Chem. Eng.,* 83:157-160.
Kowal, N.E., and Pahren, H.R., 1978, "Wastewater treatment: health effects associated with wastewater treatment and disposal," *J. Water Pollut. Contr. Fed.,* 50:1193.
Lee, D.D., Scott, C.D., and Hancher, C.W., 1979, "Fluidized bed bioreactor for coal conversion effluents," *J. Water Pollut. Contr. Fed.,* 51:974-984.
Lehtoma, K.L., and Niemela, S., 1975, "Improving microbial degradation of oil in soil," *Ambio,* 4:126.
McKenna, E.J., and Heath, R.D., 1976, "Biodegradation of polynuclear aromatic hydrocarbon pollutants by soil and water microorganisms," Univ. of Illinois Report No. UTLU-WRC-76-0113.
Munnecke, D.M., 1979, "Chemical, physical, and biological methods for the disposal and detoxification of pesticides," *Residue Review,* 70:1-26.

Nelson, R.F., and Siegrist, T.W., 1979, "Industrial wastes: chemicals and allied products," *J. Water Pollut. Contr. Fed.,* 51:1419.

Orndorff, S.A., and Colwell, R.R., 1980, "Microbial transformation of kepone," *Appl. Environ. Microbiol.,* 39:398-406.

Patrick, F.M., and Loutit, M., 1976, "Passage of metals in effluents through bacteria to higher organisms," *Water Res.,* 10:333.

Prensner, D.S., Muchmore, C.B., Gilmore, R.A., and Qazi, A.N., 1976, "Waste-water treatment by heated rotating biological discs," *Biotechnol. Bioeng.,* 18:1615.

Reese, E.T., 1977, "Degradation of polymeric carbohydrates by microbial enzymes," *Recent Adv. Phytochem.,* 11:311-367.

Robichaux, T.J., and Myrick, H.N., 1972, "Chemical enhancement of the biodegradation of crude oil pollutants," *J. Petrol. Technol.,* 24:16-20.

Suzuki, T., 1977, "Metabolism of pentachlorophenol by a soil microbe," *J. Environ. Sci. Health,* 312:113-127.

Walker, J.D., and Colwell, R.R., 1976, "Enumeration of petroleum-degrading microorganisms," *Appl. Environ. Microbiol.,* 31:198-207.

Walker, J.D., and Colwell, R.R., 1976, "Oil, mercury, and bacterial interactions," *Environ. Sci. Technol.,* 10:1145.

Watkinson, R.J., 1978, *Developments in Biodegradation of Hydrocarbons.* Applied Science Public., Essex, England.

Yamasaki, N., Yasul, T., and Matsuska, K., 1980, "Hydrothermal decomposition of polychlorinated biphenyls," *Environ. Sci. Technol.,* 14:550.

Young, J.C., 1976, "The use of enzymes and biocatalytic additives for wastewater treatment processes," *J. Water Pollut. Contr. Fed.,* 48:1-5.

VII. RISK ASSESSMENT

Adams, A.P., and Spendlive, J.C., 1970, "Coliform aerosols emitted by sewage treatment plants," *Science,* 169:1218-1220.

Chatigny, M.A., Hatch, M.T., Wolochow, H., Adler, T., Hresko, J., Macher, J., and Besemer, D., 1979, "Studies on release and survival of biological substances," *Recomb. DNA Techn. Bull.* 2:62-68.

Crawford, G.V., and Jones, P.H., 1979, "Sampling and differentiation techniques for airborne organisms emitted from wastewater," *Water Res.* 13:393.

Elliott, L., 1980, *Walk-Through Survey Report of Eli Lilly and Co. Research Labs, Indianapolis, Indiana.* Survey Date: March 28, 1980, by the Division of Surveillance, Hazard and Evaluations and Field Studies, National Institute for Occupational Safety and Health, Cincinnati, OH.

Elliott, L., 1980, *Walk-Through Survey Report of Genentech, Inc., South San Francisco, California.* Survey Date: April 8, 1980, by the Division of Surveillance, Hazard and Evaluations and Field Studies, National Institute for Occupational Safety and Health, Cincinnati, OH.

Federal Register, September 17, 1980, "Program to assess risks of recombinant DNA research; proposed first annual update," pp. 6174-78.

Federal Register, November 21, 1980, "Guidelines for research involving recombinant DNA molecules," pp. 77384-409.

Hickey, J.L.S., and Reist, P.C., 1975, "Health significance of airborne micro-organisms from wastewater treatment processes," *J. Water Pollut. Contr. Fed.,* 47:2741.

Levy, S.B., and Marshall, B., 1979, "Survival of *E. coli* host-vector systems in the human intestinal tract," *Recomb. DNA Techni. Bull.* 2:77-80.

Levy, S.B., Marshall, B., Rowse-Eagle, D., and Onderdonk, A., 1980, "Survival of *E. Coli* host-vector systems in the mammalian intestine," *Science,* 209:391-394.

Office of Research and Safety, National Cancer Institute, and the Special Committee of Safety and Health Experts, 1979, *Laboratory Safety Monograph,* U.S. Department of Health, Education, and Welfare, Washington, D.C.

Pereira, M.R., and Benjaminson, M.A., 1975, "Broadcast of microbial aerosols by stacks of sewage treatment plants," *Public Health Reports,* 90:208.

Petrocheilou, V., and Richmond, M.H., 1977, "Absence of plasmid or *E. Coli* K12 infection among laboratory personnel engaged in R plasmid research," *Gene,* 2:323-327.

Pike, R.M., 1976, "Laboratory-associated infections: summary and analysis of 3,921 cases," *Health Lab. Sci.,* 13:1-47.

Pike, R.M., 1978, "Past and present hazards of working with infectious agents." *Arch. Pathol. Lab. Med.,* 102:333-336.

Rosenberg, B., and Simon, L., 1979, "Recombinant DNA: have recent experiments assessed all the risks?" *Nature,* 282:773-774.

Sagik, B.P., and Sorber, C.A., 1979, "The survival of host-vector systems in domestic sewage treatment plants," *Recomb. DNA Techn. Bull.,* 2:55-61.

Selander, R.K., and Levin, B.R., 1980, "Genetic diversity and structure in *E. coli* populations," *Science,* 210:545-547.

Sorber, C.A., and Sagik, B.P., "Health effects of land application of wastewater and sludge: what are the risks?" *Water Sewage Works,* 125:82.

Spendlove, J.C., 1974, "Industrial, agricultural and municipal aerosol problems," *Devel. Industr. Microbiol.,* 15:20-27.

Spendlove, J.C., 1975, "Penetration of structures by microbial aerosols," *Devel. Industr. Microbiol.,* 16:427-435.

U.S. Environmental Protection Agency, 1981, "Industrial processes profile for environmental use (IPPEU): industrial applications of recombinant DNA technology," R-003-EPA-81, January, 1981.

Wade, N., 1980, "DNA: chapter of accidents at San Diego," *Science,* 209:1101-1102.

Walgate, R., 1980, "How safe will biobusiness be?" *Nature,* 283:126-127.

Wright, S., 1980, "Recombinant DNA policy: controlling large-scale processing," *Environment,* 22:29-33.

VIII. AGRICULTURE

Albrecht, S.L., Maier, R.J., Hanus, F.J., Russell, S.A., Emerich, D.W., and Evans, H.J., 1979, "Hydrogenase in *Rhizobium japonicum* increases nitrogen fixation by nodulated soybeans," *Science,* 203:1255-1257.

Andersen, L., Shanmugam, K.T., Lim, S.T., Csonka, L.N., Tait, R., Hennecke, H., Scott, D.B., Hom, S.S.M., Haury, J.F., Valentine, A., and Valentine, R.C., 1980, "Genetic engineering in agriculture with emphasis on nitrogen fixation," *Trends in Biochem. Sci.*, 35-39.

Andrews, T.J., and Lorimer, G.H., 1978, "Photorespiration–still unavoidable?" *FEBS Lett.*, 90(1):1-9.

Barney, G.E., 1980, "The Global 2000 Report to the President. Entering the Twenty-first Century," Vol. 2, 765 pp., The Technical Report, Council on Environmental Quality and the Department of State, Washington, D.C.

Beringer, J.E., Brewin, N., Johnson, A.W.B., Schulman, H.M., and Hopwood, D.A., 1979, "The Rhizobium–legume symbiosis," *Proc. Royal Soc. London Ser. B.*, 204:219-233.

Biesboer, D.D., and Mahlberg, P.G., 1979, "The effect of medium modification and selected precursors on sterol production by short-term callus cultures of *Euphorbia tirucalli*," *J. of Nat. Products*, 42(6):648-657.

Braun, A.C., and White, P.R., 1943, "Bacteriological sterility of tissues derived from secondary crown-gall tumors," *Phytopath.*, 33:85-100.

Braun, A.C., and Wood, H.N., 1976, "Suppression of the neoplastic state with the acquisition of specialized functions in cells, tissues, and organs of crown gall teratomas of tobacco," *Proc. Natl. Acad. Sci. U.S.A.*, 73:496-500.

Brill, W.J., 1979, "Nitrogen fixation: Basic to applied," *Amer. Sci.*, 67:458-466.

Brill, W.J., 1980, "Biochemical genetics of nitrogen fixation," *Microbiol. Reviews*, 44(3):449-467.

Broadbent, L., 1957, *Investigations of Virus Diseases of Brassica Crops*, Cambridge University Press.

Buchanan, R.A., Cull, I.M., Otey, F.H., and Russell, C.R., 1978, "Hydrocarbon and rubber-producing crops," *Econ. Bot.*, 32:131-153.

Bukovac, M.J., Moss, D.N., and Zelitch, I, 1975, "Carbon inputs," In: *Crop Productivity: Research Imperatives*, A.W.A. Brown, T.C. Byerby, M. Gibbs, and A. San Pietro, eds., Michigan Agricultural Experiment Station, East Lansing, Michigan, pp. 177-200.

Bulen, W.A., and LeComte, J.R., 1966, "The nitrogenase system from *Azobacter*," *Proc. Natl. Acad. Sci. U.S.A.*, 56:979-986.

Cannon, F.C., Riedel, G.E., and Ausubel, F.M., 1979, "Overlapping sequences of *Klebsiella pneumoniae* nif DNA cloned and characterized," *Mol. Gen. Genet.*, 174:59-66.

Chaleff, R.S., and Parsons, M.F., 1978, "Direct selection *in vitro* for herbicide-resistant mutants of *Nicotiana tabacum*," *Proc. Natl. Acad. Sci. U.S.A.*, 75(10):5104-5107.

Chemical Week, 1980, "Gene transfers are aimed at self-fertilizing grain," 127(21): 44-45.

Croughen, T.P., Rains, D.W., and Stararek, S.J., 1978, "Salt tolerant lines of cultured alfalfa cells," *Crop Sci.*, 18:959-963.

Decker, J.P., 1957, "Further evidence of increased carbon dioxide production accompanying photosynthesis," *J. Solar Energy Sci. Eng.* I:30-33, 1957.

Dell-Chilton, M., Drummond, H.J., Merlo, D.J., Sciaky, D., Montoya, A.L., Gordon, M.P., and Nester, E.W., 1977, "Stable incorporation of plasmid DNA into higher plant cells: The molecular basis of crown gall tumorgenesis," *Cell*, 11:263-271.

Depicker, A., Van Montagu, M., and Schell, J., 1978, "Homologous DNA sequences in different Ti-plasmids are essential for oncogenicity," *Nature (London)*, 275:150-152.

Dix, P.S., and Street, H.E., 1975, "Sodium-resistant cultured cells from *Nicotiana sylvestris* and *Capsicum annum*. *Plant Sci. Lett.*, 5:231-237.

Doggett, H., 1970, *Sorghum*, Longmans Publishing Co., London, 419 pp.

Eady, R.R., Smith, B.E., Cook, K.A., and Postgate, J.R., 1972, "Nitrogenase of *Klebsiella pneumoniae*–purification and properties of the component proteins," *Biochem. J.*, 128:655-675.

Elmerich, C., Houmard, J., Sibold, L., Manheimer, I., and Charpin, N., 1978, "Genetic and biochemical analysis of mutants induced by bacteriophage Mu DNA integration into *Klebsiella pneumoniae* nitrogen fixation genes," *Mol. Gen. Genet.*, 165:181-189.

Epstein, E., 1977, "Genetic potentials for solving problems of soil mineral stress: adaptation of crops to salinity," In: *Plant Adaptation to Mineral Stress in Problem Soils*, M.J. Wright, ed., Cornell University Press, Utica, New York, pp. 73-82.

Epstein, E., and Norlyn, J.D., 1977, "Sea-water-based crop production: A feasibility study," *Science*, 197:249-251.

Epstein, E., Norlyn, J.D., Rush, D.W., Kingsbury, R.W., Kelley, D.B., Cunningham, G.A., and Wrona, A.F., 1980, "Saline culture of crops: A genetic approach," *Science*, 210(24):399-404.

Fleming, H., and Haselkorn, R., 1973, "Differentiation in *Nostoc muscorum*: Nitrogenase is synthesized in heterocysts," *Proc. Natl. Acad. Sci. U.S.A.*, 70:2727-2731.

Flowers, T.J., Troke, P.F., and Yeo, A.R., 1977, "The mechanism of salt tolerance in halophytes," *Ann. Rev. Plant Physiol.*, 28:89-121.

Fred, E.B., Baldwin, I.L., and McCoy, E., 1932, *Root Nodule Bacteria and Leguminous Plants*, University of Wisconsin Press, Madison, WI.

Gengenbach, B.G., Green, C.E., and Donovan, C.M., 1977, "Inheritance of selected pathotoxin resistance in maize plants regenerated from cell cultures," *Proc. Natl. Acad. Sci. U.S.A.*, 74:5113-5117.

Gordon, J.K., and Brill, W.J., 1972, "Mutants that produce nitrogenase in the presence of ammonia," *Proc. Natl. Acad. Sci. U.S.A.*, 69(12):3501-3503.

Groet, S.S., and Kidd, G.H., 1980, "Embryogenesis in callus and suspension cultures of milkweed," *Plant Physiol. Suppl.*, 65:36.

Gwynne, P., 1980, "The cloning of Russet Burbank," *Mosaic*, 11(3):33-38.

Hageman, R.V., and Burris, R.H., 1978, "Nitrogenase and nitrogenase reductase associate and dissociate with each catalytic cycle," *Proc. Natl. Acad. Sci. U.S.A.*, 75:2699-2702.

Hardy, R.W.F., Filner, P., and Hageman, R.H., 1975, "Nitrogen input," In: *Crop Productivity: Research Imperatives*, A.W.A. Brown, T.C. Byerly, M. Gibbs, and A. SanPietro, eds., Michigan Agricultural Experiment Station, East Lansing, Michigan, pp. 133-176.

Hardy, R.W.F., and Havelka, V.D., 1975, "Nitrogen fixation research: A key to world food?" *Science*, 188:633-643.

Hatch, M.D., and Slack, C.R., 1970, "Photosynthetic carbon dioxide–fixation pathways," *Ann. Rev. Plant Physiol.*, 21:141-162.

Holsters, M., Silva, B., VanViet, S., Genetello, C., DeBlock, M., Dhaese, P., DePicker, A., Inze, D., Engler, G., Villarroel, R., Van Montague, M., and Schell, J., 1980, "The functional organization of the nopaline A. tumefaciens plasmid pTiC58," Plasmid, 3:212-230.

Jackson, W.A., and Volk, R.J., 1970, "Photorespiration," Ann. Rev. Plant Physiol., 21:385-432.

Kado, C.I., 1979, "Host-vector systems for genetic engineering of higher plant cells," In: Genetic Engineering–Principles and Methods, J.K. Setlow and A. Hollaender, eds., Plenum Press, New York, New York, pp. 223-239.

Kerr, A., Manigault, P., and Tempe, J., 1977, "Transfer of virulence in vivo and in vitro in Agrobacterium," Nature (London), 265:560-561.

Krol, A.J.M., Hontelez, J.G.J., Van den Bos, R.C., and Van Kammen, A., 1980, "Expression of large plasmids in the endosymbiotic form of Rhizobium leguminosarum," Nucleic Acids Res., 8(19):4337-4347.

Laetsch, W.M., 1974, "The C4 syndrome: A structural analysis," Ann. Rev. Plant Physiol., 25:27-52.

Lebeurier, G., Hirth, L., Hohn, T., and Hohn, B., 1980, "Cauliflower mosaic virus DNA passaged through bacteria remains infectious for plants," Experientia, 36:1241.

Lim, S.T., 1978, "Determination of hydrogenase in free-living cultures of Rhizobium japonicum and energy efficiency of soybean nodules," Plant Physiol., 62:609-611.

Lim, S.T., Andersen, K., Tait, R., and Valentine, R.C., 1980, "Genetic engineering in agriculture: Hydrogen uptake (hup) genes," Trends in Biochem. Sci., 5(6):67-70.

Lippincott, J.A., and Lippincott, B.B., 1975, "The genus Agrobacterium and plant tumorigenesis," Ann. Rev. Microbiol., 29:377-405.

Ljones, T., and Burris, R.H., 1978, "Evidence for one electron transfer by the Fe protein of nitrogenase," Biochem. Biophys. Res. Commun., 80:22-25.

Loomis, R.S., Williams, W.A., and Hall, A.E., 1971, "Agricultural productivity," Ann. Rev. Plant Physiol., 22:431-468.

Lung, M.C.Y., and Pirone, T.P., 1972, "Datura stramonium, a local lesion host for certain isolates of cauliflower mosaic virus," Phytopath., 62:1473-1474.

MacNeil, D., and Brill, W.J., 1978, "6-Cyanopurine, a color indicator useful for isolating mutations in the nif (nitrogen fixation) genes of Klebsiella pneumoniae," J. Bacteriol., 136:247-252.

Marx, J., 1979, "Plants: Can they live in salt water and like it?" Science 206: 1168-1169.

Massoud, F.E., 1974, "Salinity and alkalinity as soil degradation hazards," FAO/UNEP Expert Consultation on Soil Degradation, FAO, Rome.

Meagher, R.B., Shepherd, R.J., and Boyer, H.W., 1977, "The structure of cauliflower mosaic virus: I.A. Restriction endonuclease map of cauliflower mosaic virus DNA," Virol., 80:362-375.

Measures, J.C., 1975, "Role of amino acids in osmoregulation of non-halophytic bacteria," Nature (London), 257:398-400.

Mellor, J.W., 1979, "World food strategy for the 1980's–context, objectives, and approach," In: Proceedings of the International Conference on Agricultural Production: Research and Development Strategies for the 1980's, Bonn, Federal Republic of Germany, October 8-12.

Merrick, M., Filser, M., Kennedy, C., and Dixon, R., 1978, "Polarity of mutations induced by insertion of transposons Tn5, Tn7, and Tn10 into the nif gene cluster of *Klebsiella pneumoniae,*" *Mol. Gen. Genet.,* 165:103-111.

Minchin, P.E.H., and Troughton, J.H., 1980, "Quantitative interpretation of phloem translocation data," *Ann. Rev. Plant Physiol.,* 31:191-215.

Mortenson, L.E., Morris, J.A., and Yeng, D.Y., 1967, "Purification, metal composition, and properties of molybdo-ferredoxin and azoferredoxin, two of the components of the nitrogen fixing system of *Clostridium pasteurianum,*" *Biochem. Biophys. Acta,* 141:516-522.

Nasyrov, Y.S., 1978, "Genetic control of photosynthesis and improving of crop productivity," *Ann. Rev. Plant Physiol.,* 29:215-237.

National Academy of Sciences, 1975, "Agricultural production efficiency," Washington, D.C.

National Institutes of Health, July 29, 1980, "Recombinant DNA Research Actions under Guidelines," *Fed. Reg.,* 45(147):50524-50531.

National Institutes of Health, November 21, 1980, "Guidelines for Research Involving Recombinant DNA Molecules," *Fed. Reg.,* 45(227):77372-77409.

National Institutes of Health, March 12, 1981, "Recombinant DNA Research, Actions under Guidelines," *Fed. Reg.,* 46(48):16452-16457.

National Science Foundation, 1980, *The Five-Year Outlook: Problems, Opportunities, and Constraints in Science and Technology,* Volume I, U.S. Government Printing Office.

Nickell, L.G., 1977, "Crop improvement in sugarcane: Studies using *in vitro* methods," *Crop Sci.,* 17:717-719.

Nieva-Gomez, D., Roberts, G.P., Klevickis, S., and Brill, W.J., 1980, "Electron transport to nitrogenase in *Klebsiella pneumoniae,*" *Proc. Natl. Acad. Sci. U.S.A.,* 77:2555-2558.

Office of Biosafety, U.S. Department of Health, Education, and Welfare, 1974, *Classification of Etiologic Agents on the Basis of Hazard,* Fourth Edition, Center for Disease Control, Atlanta, Georgia.

Orme-Johnson, W.H., Hamilton, W.D., Ljones, T., Tso, M.Y.W., Burris, R.H., Shah, V.K., and Brill, W.J., 1972, "Electron paramagnetic resonance of nitrogenase and nitrogenase components of *Clostridium pasteurianum* W5 and *Azotobacteria vinelandii,*" *Proc. Natl. Acad. Sci. U.S.A.,* 69:3142-3145.

Oswald, T.H., Smith, A.E., and Phillips, D.V., 1977, "Herbicide tolerance developed in cell suspension cultures of perennial white clover," *Can. J. Bot.,* 55:1351-1358.

Paul, D.A., Kilmer, R.L., Altobello, M.A., and Harrington, D.N., 1977, *The Changing U.S. Fertilizer Industry,* U.S. Department of Agriculture, Economic Research Service, Agricultural Economic Report No. 378, 103 pp.

Peters, G.A., 1978, "Blue-green algae and algal associations," *BioSci.,* 28:580-585.

Peterson, R.B., and Wolk, C.P., 1978, "High recovery of nitrogenase activity and of [55]Fe-labeled nitrogenase in heterocysts isolated from *Anabaena variabilis,*" *Proc. Natl. Acad. Sci. U.S.A.,* 75:6271-6275.

Phillips, D.A., and M.J. Johnson, 1961, "Aeration in fermentations," *J. Biochem. Microbiol. Technol. Eng.,* 111:227-309.

Ponnamperuma, F.N., 1977, In: *Plant Response to Salinity and Water Stress,* W.J.S. Downton, and M.G. Pitman, eds., Association for Scientific Cooperation in Asia, Sydney.

Postgate, J.R., 1974, "New advances and future potential in biological nitrogen fixation," *J. of Appl. Bacteriol.,* 37:185-202.

Postgate, J.R., 1977, "Consequences of the transfer of nitrogen fixation genes to new hosts," *Ambio.,* 6(2/3):178-180.

Princen, L.H., 1977, "Potential wealth in new crops," In: *Crop Resources,* D.S. Seigler, ed., Academic Press, New York, pp. 1-16.

Radin, D.N., and Carlson, P.S., 1978, "Herbicide tolerant tobacco mutants selected *in situ* and recovered in regeneration from cell culture," *Genet. Res. (Camb.),* 32:85-89.

Rains, D.W., 1979, "Salt tolerance of plants: Strategies of biological systems," In: *The Biosaline Concept,* A. Hollaender, ed., Plenum Press, NY., pp. 47-68.

Riedel, G.E., Ausubel, F.M., and Cannon, F.C., 1979, "Physical map of chromosomal nitrogen fixation (nif) genes of *Klebsiella pneumoniae,*" *Proc. Natl. Acad. Sci. U.S.A.,* 76:2866-2870.

Roberts, G.P., MacNeil, T., MacNeil, D., and Brill, W.J., 1978, "Regulation and characterization of protein products coded by the nif (nitrogen fixation) genes of *Klebsiella pneumoniae,*" *J. Bacteriol.,* 163:267-279.

Saterson, K.A., Luppold, M.K., Scow, K.M., and Lee, R.E., 1979, "Herbaceous Species Screening Program Phase I," Final Report, U.S. Department of Energy, Biomass Energy Systems Branch, Washington, D.C., 132 pp. (Arthur D. Little, Inc., Cambridge, Massachusetts).

Schell, J. and Van Montagu, M., 1978, "Transfer, maintenance, and expression of bacterial Ti-plasmid DNA in plant cells transformed with *Agrobacterium tumefaciens,*" *Brookhaven Symp. Biol.,* 29:36-49.

Schell, J., Van Montagu, M., DeBeuckeleer, M., DeBlock, M., Depicker, A., DeWilde, M., Engler, G., Genetello, C., Hernalsteens, J.P., Holsters, M., Seurinck, J., Silva, B., Van Vliet, F., and Villarroel, R., 1979, "Interactions and DNA transfer between *Agrobacterium tumefaciens,* the Ti-plasmid and the plant host," *Proc. Royal Soc. London Ser. B.,* 204:251-266.

Schubert, E.R., and Evans, M.J., 1976, "Hydrogen-evolution: A major factor affecting the efficiency of nitrogen fixation in nodulated symbionts," *Proc. Natl. Acad. Sci. U.S.A.,* 73:1207-1211.

Schubert, E.R., and Evans, M.J., 1977, In: *Recent Developments in Nitrogen Fixation,* W. Newton, J.R. Postgate, and C. Rodeniguez-Barreuco, eds., Academic Press, NY, pp 471-485.

Shah, V.K., and Brill, W.J., 1973, "Nitrogenase IV. Simple method of purification to homogeneity of nitrogenase components from *Azotobacter vinelandii,*" *Biochem. Biophys. Acta,* 305:445-454.

Shanmugam, K.J., O'Gara, F., Andersen, K., and Valentine, R.C., 1978, "Biological nitrogen fixation," *Ann. Rev. Plant Physiol.,* 29:263-276.

Shepard, J.F., Bidney, D., and Shahin, E., 1980, "Potato protoplasts in crop improvement," *Science,* 208:17-24.

Shepherd, R.J., 1976, "DNA viruses of higher plants," *Adv. Virus. Res.,* 20:305-339.

Shepherd, R.J., Bruening, G.E., and Wakeman, R.J., 1970, "Double-stranded DNA from cauliflower mosaic virus," *Virology,* 41:339-347.

Smith, B.E., Lowe, D.J., and Bray, R.C., 1972, "Nitrogenase of *Klebsiella pneumoniae:* Electron paramagnetic resonance studies of the catalytic mechanism," *Biochem. J.,* 130:641-643.

Smith, B.E., Lowe, D.J., and Bray, R.C., 1973, "Studies by electron paramagnetic resonance on the catalytic mechanism of nitrogenase of *Klebsiella pneumoniae, Biochem. J.,* 135:331-341.

St. John, R.T., Johnston, H.M., Seidman, C., Garfinkel, D., Gordon, J.F., Shah, V.K., and Brill, W.J., 1975, "Biochemistry and genetics of *Klebsiella pneumoniae* mutant strains unable to fix nitrogen," *J. Bacteriol.* 121:759-765.

Szeto, W.W., Hamer, D.H., Carlson, D.S., and Thomas, C.A., 1977, "Cloning of cauliflower mosaic virus (CLMV) DNA in *Escherichia coli,"* *Science,* 196:210-212.

Tel-Or, E., and Stewart, W.D.P., 1977, "Photosynthetic components and activities of nitrogen-fixing isolated heterocysts of *Anabaena cylindrica,"* *Proc. Royal Soc. London, Ser. B.,* 198:61-86.

Thorpe, T.A., 1978, *Frontiers of Plant Tissue Culture,* The International Association for Plant Tissue Culture, University of Calgary, Calgary, Alberta, Canada, 556 pp.

Torrey, J.G., 1978, "Nitrogen fixation by actinomycete-nodulated angiosperms," *BioSci.,* 28:586-592.

U.S. Department of Agriculture, Economics and Statistics Service, 1980, 1981, Fertilizer Situation, 33 pp.

Vandercasteele, J.P., and Burris, R.H., 1970, "Purification and properties of the constituents of the nitrogenase complex from *Clostridium pasteurianum,"* *J. Bacteriol.,* 101:794-801.

Velthuys, B.R., 1980, "Mechanisms of electron flow in photosystem II and toward photosystem I," *Ann. Rev. of Plant Physiol.,* 31:545-567.

Vincent, J.M., 1974, "Root-nodule symbioses with *Rhizobium,"* In: *The Biology of Nitrogen Fixation,* A. Quispel, ed., North Holland Publishing Company, Amsterdam, Holland, pp. 265-341.

White, F.F., and Nester, E.W., 1980, "Hairy root: Plasmid encodes virulence traits in *Agrobacterium rhizogenes,"* *J. Bacteriol.,* 141:1134-1141.

Whittaker, R.H., and Feeny, P.P., 1971, "Allelochemics: Chemical interactions between species," *Science,* 171:757-770.

Williams, M.C., 1980, "Purposefully introduced plants that have become noxious or poisonous weeds," *Weed Sci.,* 28(3):300-305.

Yadav, N.S., Postle, K., Saiki, R.K., Thomashow, M.F., and Chilton, M.D., 1980, "T-DNA of a crown gall teratoma is covalently joined to host plant DNA," *Nature (London),* 287:458-461.

Yang, F., Montoya, A.L., Merlo, D.J., Drummond, M.H., Chilton, M.D., Nester, E.W., and Gordon, M.P., 1980, "Foreign DNA sequences in crown gall teratoma and their fate during the loss of the tumorous traits," *Mol. Gen. Genet.,* 177:707-714.

Yates, M.G., 1971, "Electron transport to nitrogenase in *Azotobacter chroococcum,* Purification and some properties of NADH dehydrogenase," *Eur. J. Biochem.,* 24:347-357.

Yoch, D.C., 1974, "Electron transport carriers involved in nitrogen fixation by the coliform *Klebsiella pneumoniae,*" *J. Gen. Microbiol.,* 83:153-164.

Zaenen, I., Van Larebeke, N., Teuchy, H., Van Montagu, M., and Schell, J., 1974, "Supercoiled circular DNA in crown-gall inducing *Agrobacterium* strains," *J. Mol. Biol.,* 86:109-127.

Zelitch, I., 1979, "Photosynthesis and plant productivity," *Chem. and Eng. News,* 5(57):28-48.

Index

187

Other Noyes Publications

GENETIC ENGINEERING APPLICATIONS FOR INDUSTRY

Edited by J.K. Paul

Chemical Technology Review No. 197

The documents that make up this book are four working papers which were prepared by *Genex Corporation, Poly-Planning Services,* and the *Massachusetts Institute of Technology* for the U.S. Congress's Office of Technology Assessment (OTA) report on the impacts of genetic engineering. **These papers contain significantly more technical information than is found in the OTA report.** They are published in their entirety.

Genetic engineering, with all of its implications and ramifications, is the most exciting area of scientific research in the world today.

Applications of genetic engineering offer almost unlimited challenges in such industries as chemicals, foods, pharmaceuticals, agricultural products, and energy, and, of course, in medical research. The book includes technical discussions of previous accomplishments and potential research directions for the future.

A **condensed table of contents, with selected subtitles,** is given below.

ISBN 0-8155-0869-7

580 pages

Other Noyes Publications

INDUSTRIAL ENZYMES FROM MICROBIAL SOURCES
Recent Advances

Edited by M.G. Halpern

Chemical Technology Review No. 186

The production of a wide variety of industrially valuable enzymes from microbial sources, such as bacteria, fungi and yeasts, is described. Nearly 200 processes detail the enzyme production methods.

The demand for industrial enzymes, both in the U.S. and Europe, is growing rapidly and the near term prospects appear to be excellent. Enzyme-produced products currently in great demand include cornstarch-derived sweeteners and motor-fuel-grade alcohol, but the future of this field seems wide open, especially as genetic engineering becomes an accepted tool.

Potential uses for these enzymes can be found in such commercial applications as detergent formulation, food product manufacture, pharmaceutical manufacture, chemical manufacture, and in the preparation of medical diagnostic materials.

Below is a condensed table of contents including **chapter titles and selected subtitles.** The number of processes in any chapter is shown in parentheses.

ISBN 0-8155-0843-3 (1981)

346 pages